DIMENSIONAL VARIATION MANAGEMENT HANDBOOK

DIMENSIONAL VARIATION MANAGEMENT HANDBOOK

A Guide for Quality, Design, and Manufacturing Engineers

John V. Liggett
President
Variation Management Consultants
Northville, MI

Prentice Hall
Englewood Cliffs, New Jersey 07632

Library of Congress Cataloging-in-Publication Data

Liggett, John V.
Dimensional variation management handbook : a guide for quality, design, and manufacturing engineers / John V. Liggett.
p. cm.
Rev. ed. of: Fundamentals of position tolerance / John V. Liggett, 1970.
Includes index.
ISBN 0–13–927641–6
1. Tolerance (Engineering)—Handbooks, manuals, etc. I. Liggett, John V. Fundamentals of position tolerance. II. Title.
TS172.L47 1993 92–28359
620′.0045—dc20 CIP

Acquisition Editor: Mike Hays
Production Editors: Fred Dahl and Rose Kernan
Copy Editor: Rose Kernan
Designers: Fred Dahl and Rose Kernan
Cover Designer: Ben Santora
Prepress Buyer: Mary McCartney
Manufacturing Buyer: Susan Brunke

Printed in the United States of America
10 9 8 7 6 5 4 3 2 1

ISBN 0-13-927641-6

Prentice-Hall International (UK) Limited, *London*
Prentice-Hall of Australia Pty. Limited, *Sydney*
Prentice-Hall Canada Inc., *Toronto*
Prentice-Hall Hispanoamericana, S.A., *Mexico*
Prentice-Hall of India Private Limited, *New Delhi*
Prentice-Hall of Japan, Inc., *Tokyo*
Simon & Schuster Asia Pte. Ltd., *Singapore*
Editora Prentice-Hall do Brasil, Ltda., *Rio de Janeiro*

Dedicated to my sons Mike, Pat, and John

Contents

CHAPTER 3

Statistical Analysis, 23

CHAPTER 4

Statistics in Manufacturing, 49

CHAPTER 5

Introduction to Geometric Tolerancing, 65

CHAPTER 6

Datums, 80

CHAPTER 7

Geometric Tolerances, 102

CHAPTER 8

The Boundary Concept, 126

CHAPTER 9

Fundamentals of Position Tolerance, 131

CHAPTER 10

Applications of Position Tolerance, 153

Preface

Dimensional Variation Management Handbook is an expansion of my now out-of-print book *Fundamentals of Position Tolerance* and a course that I taught at the University of Michigan. That course, entitled *Tolerance Considerations in Design,* was my first attempt at expanding Position Tolerance theory to include geometric tolerances and to create a broader, analytical-based methodology. This handbook is another step in this evolution toward a broad perspective which considers all aspects of dimensional variation of manufactured products and the variety of techniques used to analyze and manage such variability.

Recent broad interest in world class manufacturing has resulted in a heightened consciousness of the need to manage the variability of manufactured products. Simultaneous engineering brings together the process and product design considerations at an early point in such programs. An important task of such efforts is to control product variability and hence, product quality, through planning, analysis, and discussion of the variability introduced by the various manufacturing processes. Design and process tradeoffs made early in such programs can have a major effect upon product variability, subsequent product quality, process reliability, and cost.

Many recent trends have heightened the engineering and manufacturing community's interest and ability to manage variability in a comprehensive manner. Other factors have a smaller but significant effect (e.g., the dominant use of symbols to communicate tolerance information). Major influences on this book include:

1. Broader use of statistical methods in dimensional analysis and manufacturing control in the United States.

2. Gradual shift to metric (SI) units of measurement in many U.S. industries. Metric dimensions (mm) have been used throughout this book.
3. Emergence of ANSI-Y14.5M-1982, *Dimensioning and Tolerancing,* as the uniform U.S. standard defining methods of tolerancing and use of tolerance symbols. Although Y14.5 is a mature document, the reader will find that I make several suggestions that I trust ultimately will change this standard. The reason for these suggestions are that the standard developed from an application base of rigid prismatic parts. Application to automotive body structures and other similar parts highlights the need for more comprehensive methods that will work well with:
   a) Non-rigid parts.
   b) Complex three-dimensional surface shapes.
   c) Computer-Aided Designs (CAD) based upon a math data base rather than a dimensional pattern.

   These suggestions are directed toward removing the restrictions resulting from definitions and examples based upon rigid prismatic parts thereby creating a standard with broader applicability.
4. My many friends at Fetz Engineering in Sterling Heights, Michigan. Fetz is a leading variation management engineering firm that works with automotive body structures. I spent several months working with the Fetz organization, getting current on industry trends while I completed this book.

Special thanks also go to Dick Cumming of the Cadillac Division of GM for the time he spent reading my preliminary manuscript for Prentice Hall and Randy Dawley of General Dynamics Land Systems Division who proofread the final manuscript. Both offered many helpful suggestions. Melissa Manhart did a wonderful job of turning my sketches into the high quality computer-generated illustrations used throughout the book.

John V. Liggett
Northville, Michigan

Introduction

This book is about the dimensional variation of manufactured products and how design, manufacturing, and quality engineers manage it. Manufactured products often appear to be identical because the variation is small and those involved do not recognize the variability that can be shown by precise instruments. We all know from our own experience that no two things are exactly the same. People's height, the shape of snow flakes, and our finger prints are variables that come easily to mind. When we extend this realization it is also easy to recognize that no two manufactured products are exactly the same. Once we have consciously accepted that such variation exists, we can then go about the business of managing it. Hence, the discipline of variation management.

A subject is more easily understood if the philosophies that drive it are known and understood. In the case of *variation management,* the principal motivator is the broad interest among manufacturing organizations in achieving world class capability and status for their manufacturing operations and products. Quality philosophies have matured slowly, being driven in the last quarter century by American-derived, Japanese-refined, statistical methodology. *Zero defect concepts* such as the Poka-yoke methods of Shingo[1] are also significant, as they may supplement and potentially surpass their statistical counter parts. Variation management, a cornerstone of any world class manufacturing approach, seeks to understand the sources of variation, select product features and manufacturing processes that minimize variation, and use appropriate methods to communicate the expected level of variation and the designated means to control it.

[1] Shingo, Shigeo. *Zero Quality Control: Source Inspection and the Poka-yoke System,* Productivity Press, 1986.

Some of the building blocks of variation management are:

1. Analytical methods which can range from linear stack-ups to process charts to statistical stack-ups, sometimes referred to as RSS (from *root sum square*). Modern computers, which surpass the capability of mainframe units of only a few years ago, permit the use of simulation techniques to apply sophisticated methodology to complex system analysis. MCS, or *Monte Carlo simulation* is one of the current areas of keen interest in variation management.
2. Methods of communication which range from complex control documents (such as those used to define and control an automobile body), to tolerance symbols used for notation which is frequently referred to as GD&T from *geometric dimensioning and tolerancing,* which I now refer to as *geometric datums and tolerances.*
3. Manufacturing methods that are driven by the datums and tolerance values contained in the engineering specifications. The magnitude of specific tolerances can dictate the process or machinery used to produce a subject part. Datum requirements dictate fixture location points and relationships between operations. Properly read, GD&T requirements and *critical characteristic* notations totally define most gages.
4. Recognition that control of variation is a management responsibility that requires diverse inputs, such as those that result from the interaction of design and manufacturing engineers engaged in simultaneous engineering programs. Although much of the responsibility for variation management lies with the product engineering community, more contemporary organizations recognize that the manufacturing community also has a significant role in this endeavor. Examples include complex tolerances that allow tradeoffs, such as with position and size of holes, and statistical methods to develop process controls and define process capability.

The principal goal of this book is to provide a thorough understanding of the many faces of variation management and thus an in-depth appreciation of an overall management approach to controlling the elements of variation of manufactured products. It is also our intent to provide sufficient instruction in specific techniques so that design, manufacturing, and quality engineers will be able to calculate specific values, document results, understand the viewpoint of their colleagues, and, in general, participate in the variation management process as full and active partners.

Symbols Used

C = Concentricity or eccentricity expressed as total

C_{ADD} = Additional concentricity

C_C = Critical concentricity

C_C = Process centering parameter

C_T = Total concentricity

CL = Clearance

CL_{AV} = Average of two clearances

C_P = Process capability ratio

C_{pk} = Process centering ratio

D = Hole depth

E = Eccentricity

E_T = Total eccentricity

F = Fastener size at MMC

F_A = Actual fastener size

F_L = Fastener size at LMC

F_V = Virtual fastener size

H = Hole size at MMC
H_A = Actual hole size
H_L = Hole size at LMC
HD_L = Head size at LMC
K_B = Correction factor for balanced tolerances
K_R = Correction factor for round holes
K_S = Correction factor for square holes
L = Overlap
LMC = Least material condition
M = Movement of fastener
MMC = Maximum Material Condition
P = Probability
P = Projected height
R_F = Resultant size for fastener
F_H = Resultant size for hole
RFS = Regardless of feature size
S = Sigma (standard deviation sample population)
T = Tolerance
T_A = Allowable or actual tolerance
T_{ADD} = Additional tolerance
T_{ASSY} = Tolerance of assembled parts
T_B = Balanced position tolerance
T_C = Critical position tolerance
T_F = Assembled fastener tolerance
T_i = Tolerance of individual component of assembly
T_T = Total position tolerance for two parts
V_F = Virtual size for a fastener
V_H = Virtual size for a hole
$\overline{X}$ = Mean value of several Xs

DIMENSIONAL VARIATION MANAGEMENT HANDBOOK

Chapter 1

Variation Analysis

Product definition must always be based upon functional requirements. Like the old real estate axiom, there are three important things to consider in tolerancing a component: function, function, and function. We must also consider that manufacturing processes always induce variations, however small, which effect the dimensions and other key parameters such as: strength, electrical properties, weight, and the like. The confluence of these two thoughts gives rise to many questions about the effects of variability upon desired functions.

Analysis of the variable dimensions in an assembled product is one of the fundamental tasks in variation management. Such analysis, frequently called *stack-ups*, are conducted for many reasons. Some of these reasons are:

1. To ensure interchangeability of mating parts so as to make economical assembly processes possible and to also make possible the replacement of parts for repair or service.

2. To ensure clearance between adjacent parts not directly dimensioned or controlled at the time of assembly. Dimension A in Figure 1-1 is typical of this class of problem. Analysis of a range of conditions to ensure that available adjustments will suffice is a similar task.

3. To determine the impact of variability upon aesthetical characteristics. In an automobile body, the parallelism of one panel edge to an adjacent panel edge and the resulting straight or tapered gap, the size of gaps on opposing sides of a panel, the deviation from a desired flush position of adjacent panels are important issues.

4. To ensure that a function, usually of a secondary nature, that is not directly controlled performs adequately. For example, consider the retention of a disc brake piston

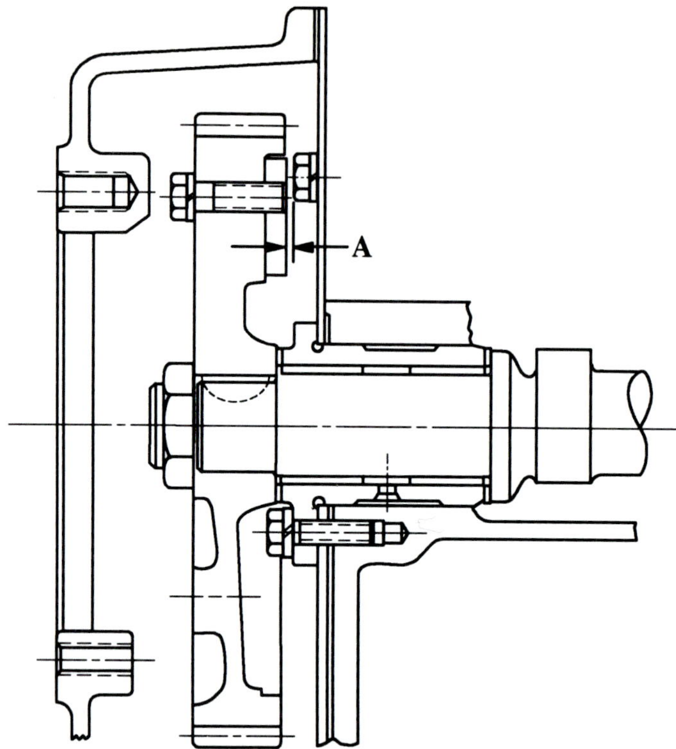

Figure 1-1. Indirectly controlled clearance

when the brake lining is fully worn (see Figure 1-2). Obviously, a serious safety issue exists in such a case, although the function is of a secondary nature, following proper application of the brake.

5. To ensure that dimensions and tolerances fit a manufacturing sequence. Assembly techniques such as selective assembly and shimming also require careful analysis so that they can be successfully implemented in the manufacturing cycle.

In this chapter, we will consider linear analysis techniques. Statistical and other more complex methods will build upon these techniques in Chapter 2. While we will use relatively simple examples to convey concepts as clearly as possible, it should be noted that real world analyses can include many elements, some of them apparently contradictory, confusing, or vague. A careful statement of objectives, distillation of available data, and application of the techniques that follow will yield sound conclusions.

VARIATION VERSUS TOLERANCE

Before we proceed further, let us talk about variation and tolerance. Variation is the change that occurs in size and shape of features or parts which we encounter. Tolerances are engineering specifications that define the extremes in variation from the ideal or nominal di-

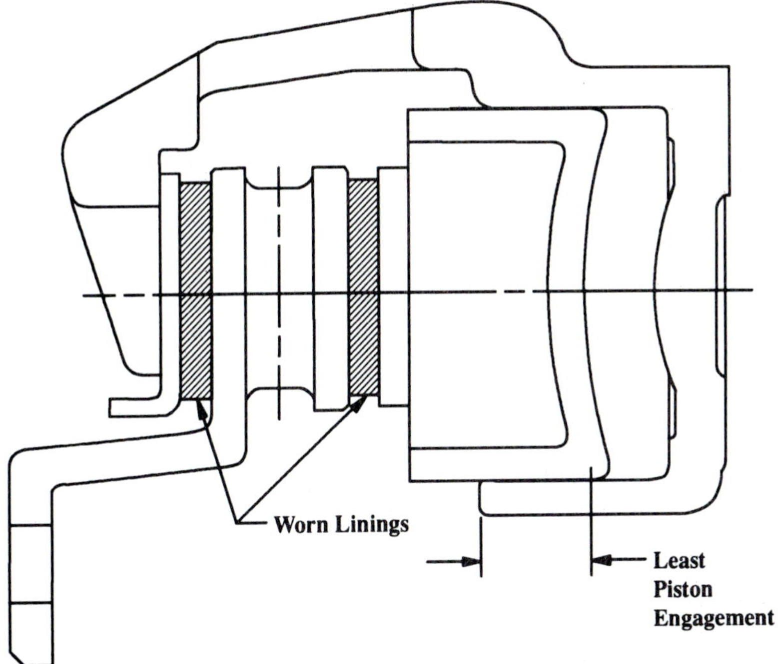

Figure 1-2. Disc brake caliper with worn linings

mension that we will allow or accept. When we do analyses, we are attempting to model the effect of component variation upon assembled units. But such variation is not known, so we use design means and tolerance values to represent the extreme variation that will occur if the product specifications are met. This means that *the tolerance values must be reasonably aligned with the process mean and spread or an analysis based upon tolerances will not be meaningful.* Table 1-1 outlines the consequence of mismatching design and process values.

Table 1-1. Tolerance vs. process capability

TOLERANCE	MANUFACTURING PROCESS	ANALYSIS PREDICTS
Significantly smaller than process	Incapable	Better results than are possible
Equal to process	Capable	True results
Significantly larger than process	Very capable but process mean shifts are likely	Results much worst than will occur

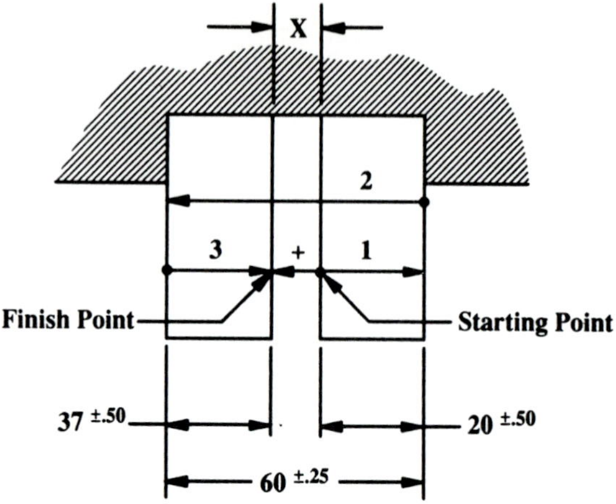

Figure 1-3. Assembly with bilateral tolerance

BILATERAL TOLERANCES

Figure 1-3 illustrates a simple example using bilateral tolerance (equal, plus, and minus values). This example has two parts that fit into a slot. We must calculate the dimension *X* between them. This may be done to ensure that a space remains or to determine how much space is available in order to add another part.

The following steps, or rules, define a method[1] to perform the calculation. Although these rules may seem rigorous for such a simple example, it is important that they be followed as many other calculations will be significantly more complex. The rules are designed to prevent errors in calculation due to variations in dimensioning patterns and tolerancing methods.

1. Establish starting and finish points.
2. Go from the starting point to the finish point in the shortest, most direct route. Call this direction +.
3. Go around the circuit in the opposite direction as shown (dimensions 1 through 3). More complex calculations may alternate back and forth in direction as in Example 1-2.
4. Add signs to the nominal values according to their direction. Mark the + direction above the calculation for reference and add the nominal values algebraically.
5. Add all tolerances.

[1]This technique (see Figure 1-1 and Example 1-2) is based upon "Tolerance Stack-Up on Drawings" by Albin Chapin, *Machine Design Magazine* July 5, 1962. Reprinted with permission.

Dimensions	*Tolerances*
+	
←	
−20	0.50
+60	0.25
−37	0.50
+3	1.25

Figure 1-3a.

$X = 3 \pm 1.25$, or expressed as a range = 1.75 to 4.25.

Note that the resulting sign is +, matching the assigned direction, thus confirming that clearance exists between the parts. If a minus sign had resulted, an interference condition would have been indicated.

It is not unusual to find a case with values ranging between clearance and interference. Such a result might look like:

1.0 ± 2.5, or expressed as a range = −1.5 to 3.5.

From −1.5 to 0 would indicate interference, while 0 to 3.5 would indicate clearance.

UNBALANCED TOLERANCES

When unbalanced tolerances (unequal + and − values) or unilateral tolerances (either the + or − tolerance is 0) are encountered, the basic procedure outlined still applies, however, the + and − tolerances must be considered separately. Figure 1-4 illustrates such an instance.

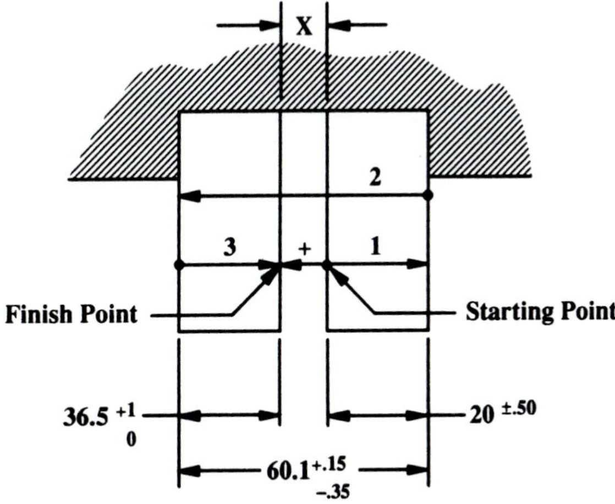

Figure 1-4. Assembly with unbalanced and unilateral

Examination of the individual dimensions shows that they have the same limits as in Figure 1-3 except that the tolerances are now expressed as unbalanced or unilateral. Again the rules are listed below. Those that differ from the previous case are shown in italics.

1. Establish starting and finish points.
2. Go from the starting point to the finish point in the shortest, most direct route. Call this direction +.
3. Go around the circuit in the opposite direction as shown (dimensions 1 through 3). More complex calculations may alternate back and forth in direction as in Example 1-2.
4. Add signs to the nominal values according to their direction. Mark the + direction above the calculation for reference and add the nominal values algebraically.
5. *Develop a sign for each tolerance based upon its original sign and its direction sign. For example, the tolerance on dimension 3 has a + value and a – direction, resulting in a – value for the analysis. CAUTION—ERRORS IN TOLERANCE SIGNS WILL INVALIDATE THE ANALYSIS.*

Dimension	*+ Tolerance*	*– Tolerance*
+ ←		
–20	.50	.50
+60.1	.15	.35
–36.5	.00	1.00
+3.6	+.65	–1.85

Figure 1-4a.

$X = 3.6 + 0.65 - 1.85$, or expressed as a range = 1.75 to 4.25 as before.

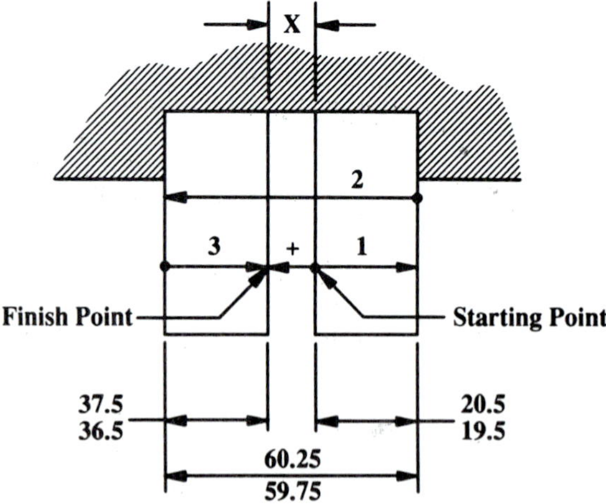

Figure 1-5. Assembly with limit dimensions

LIMIT DIMENSIONS

Figure 1-5 shows the same example with the previously toleranced nominal dimensions converted to limit dimensions. Again, the basic rules apply and those specific to this category are in italics.

1. Establish starting and finish points.
2. Go from the starting point to the finish point in the shortest, most direct route. Call this direction +.
3. Go around the circuit in the opposite direction as shown (dimensions 1 through 3). More complex calculations may alternate back and forth in direction as in Figure 1-7.
4. *Sum all maximum + direction values and all minimum – direction values.*
5. *Sum all minimum + direction values and all maximum – direction values.*

Max+/Min–	*Min+/Max–*
+ ←	
+60.25	+59.75
−19.50	−20.50
−36.50	−37.5
+4.25	+1.75

Figure 1-5a.

X = 1.75 TO 4.25, as before.

Example 1-1 and 1-2 show more complex examples of bilateral tolerance stack-ups.

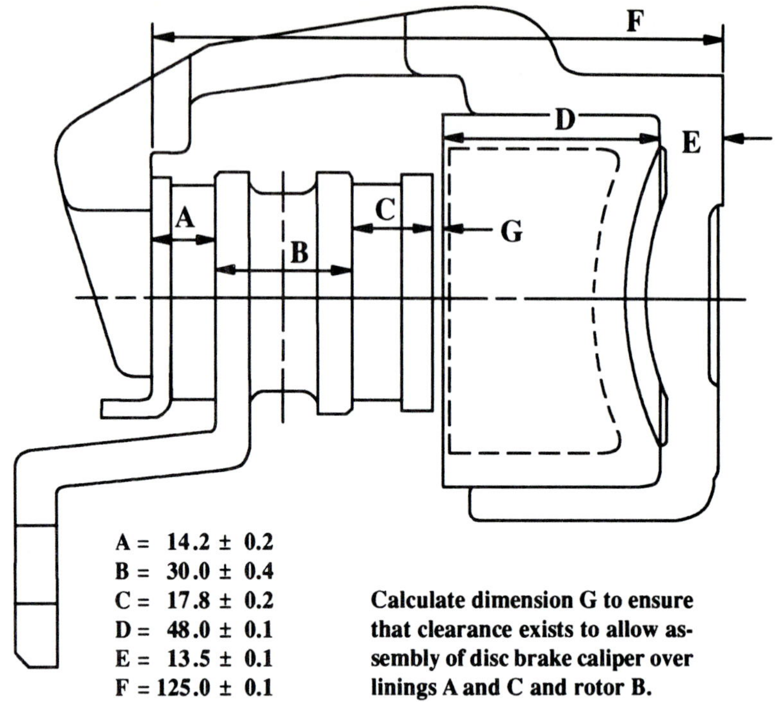

Example 1-1. Analysis for assembly clearance

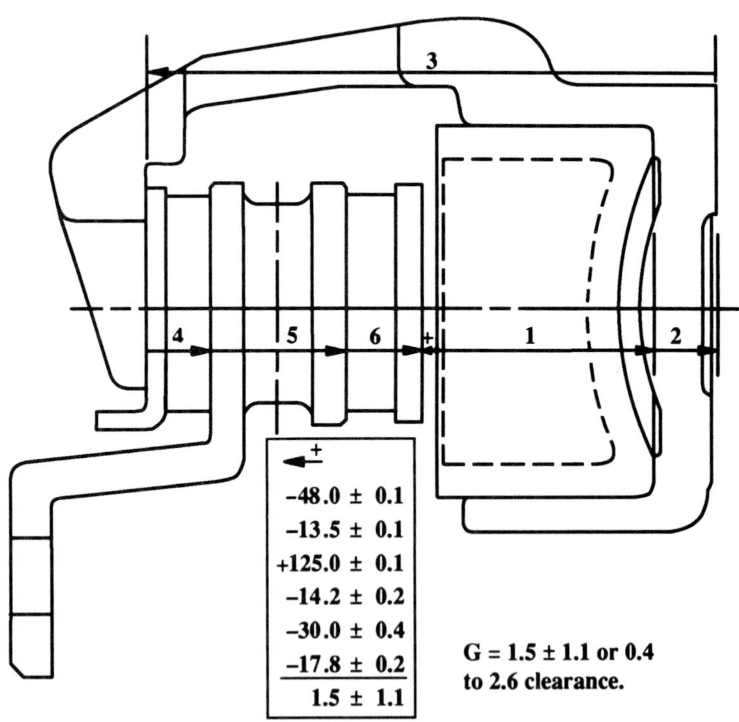

Example 1-1. Solution

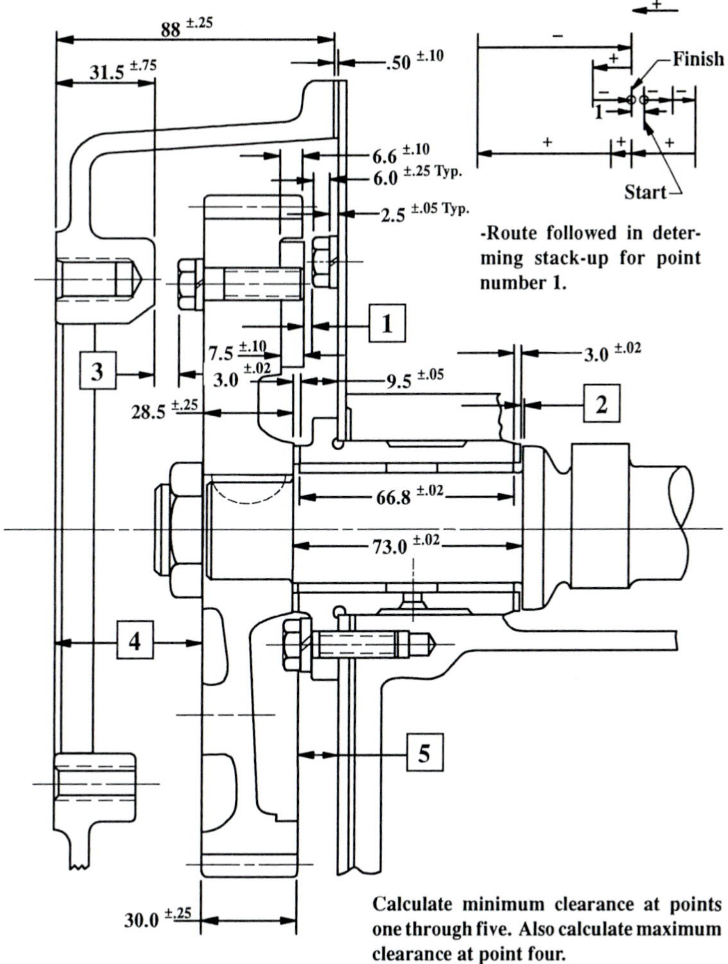

Example 1-2. Analysis for assembly clearance

1

Dim.	Tol.
+ ←	
-6.0	±.25
-2.5	.05
+9.5	.05
+3.0	.02
+28.5	.25
-30.0	.25
+6.6	.10
-7.5	.10
1.6	±1.07

Minimum

1.6 - 1.07=
0.53

2

Dim.	Tol.
+ ←	
+73.0	±.02
-3.0	.02
-66.8	.02
-3.0	.02
0.2	±.08

Minimum

0.2 - 0.08=
0.12

3

Dim.	Tol.
+ ←	
-6.0	±.25
-2.5	.05
-28.5	.25
-73.0	.02
+3.0	.02
+66.8	.02
-9.5	.05
+0.5	.10
+88.0	.25
-31.5	.75
7.3	±1.76

Minimum

7.3 - 1.76=
5.54

4

Dim.	Tol.
Gear to right	
+ ←	
-28.5	±.25
-3.0	.02
-9.5	.05
+0.5	.10
+88.0	.25
47.5	±.67
Gear to left	
-28.5	±.25
-73.0	.02
+3.0	.02
+66.8	.02
-9.5	.05
+0.5	.10
+88.0	.25
47.3	±.71

Maximum

47.5 + .67=
48.17

Minimum

47.3 - .71=
46.59

5

Dim.	Tol.
+ ←	
+9.5	±.05
+3.0	.02
+28.5	.25
-30.0	.25
11.0	±.57

Minimum

11.0 - .057=
10.43

Comments:

Points 1 and 5 are at a minimum when gear is shifted to the right. Point 3 is minimum when gear is shifted to the left.

Example 1-2. Solutions

SUMMARY

Variation analyses or stack-ups follow simple rules based upon direction signs. Careful use of sign convention is mandatory to achieve valid results. Some analysts convert all dimensions and tolerances to bilateral as a means to avoid errors in their calculations. Since bilateral tolerance are required in statistical analyses, as will be seen in later chapters, we concur in this approach and recommend it. A note of caution: When tolerances are converted, the nominal value must be shifted to a mean value (e.g., 36.5 + 1.0 – 0 becomes 37.0 ± 0.5).

In some analyses for CAD-designed products, nominal values are not stated as dimensions, being contained in the computer as a math data base. In such cases, the shift of the nominal value must be identified by other means. Preferably, the bilateral tolerance contained in the conclusion of the analysis is converted to an unbalanced tolerance as shown in Example 1-3. This allows the math data base nominal value to remain intact.

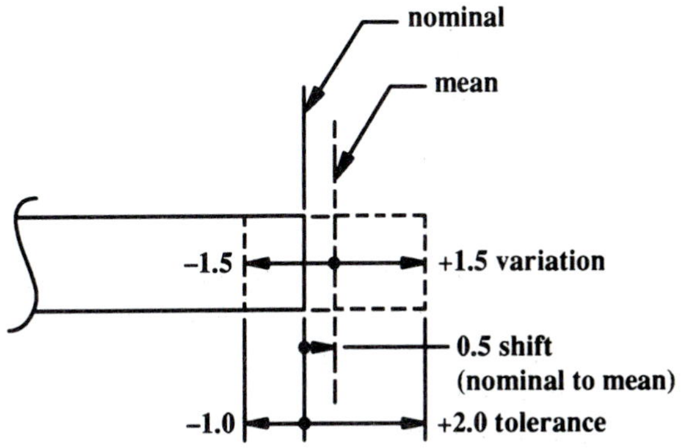

Given: **Nominal unstated**
Mean shift = +0.5
Variation = ±1.5; then
Tolerance at nominal = $^{+2.0}_{-1.0}$

Example 1-3. Tolerance conversion for mean shift

Chapter 2

Process Charts

Process charting is a variation management technique that starts with a product that has been functionally dimensioned and develops new dimensions and tolerances that reflect the manufacturing sequence. Frequently, this technique finds use on machined parts, or on castings which will subsequently be machined. Consider for example a surface that is finished in two operations to enhance its quality. Obviously, the rough cut surface with additional stock requires dimensional information not on the product drawing. Process charts can define optimum stock material required for stock removal, thereby avoiding waste and minimizing the number of machine cuts required. With optimum stock removal, unnecessarily heavy machine cuts can be avoided and tool life increased. The information developed on a *process chart* is usually communicated via a process or manufacturing instruction sheet.

Standard forms[1] and symbols are used to avoid errors as the charts quickly become complex, even when the parts are relatively simple. Figure 2-1 displays some of these notations. Dimensions and tolerances are always expressed as a mean dimension with bilateral tolerance. As noted in Chapter 1, this technique is an important method for avoiding errors when complex tolerance interactions are studied. As previously noted, unbalanced or unilateral tolerances must be converted to appropriate mean dimensions and bilateral tolerances. The part is always sketched at the top of the form to guide the analyst. A process sequence must be known or assumed to be able to prepare a process chart.

[1]The forms and examples in this chapter are based upon Cartuscello, M. A. "Tolerance Charts: Their Use and Preparation," *Production Magazine*, December 1959. Reprinted with permission.

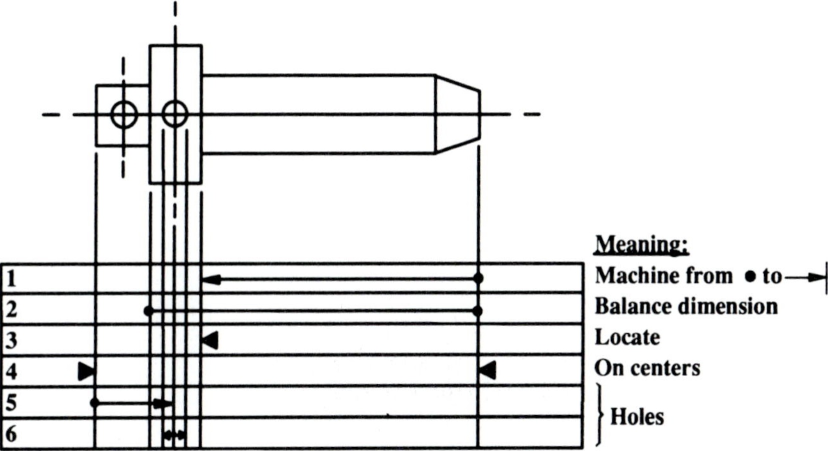

Figure 2-1. Symbols used for process charts

STOCK REMOVAL TOLERANCE

A simple part is shown in Example 2-1. Here a pin is cut off from a bar and then finish machined on the cut off end in the next operation. As seen in the example, the tolerance on the stock to be removed is the sum of the assigned rough and finish machining tolerances. A minimum allowance based upon the machine capability and the physical properties of the

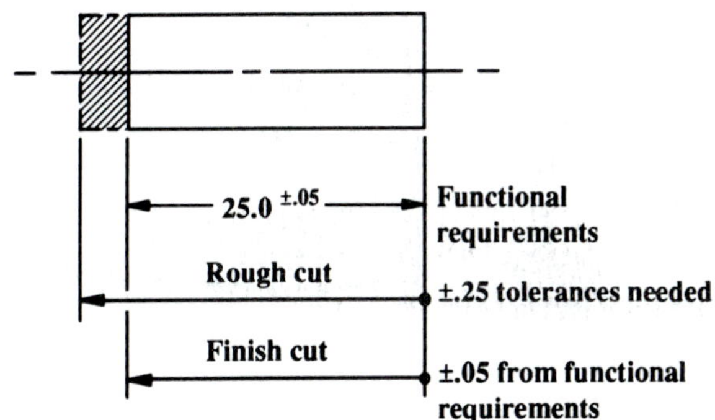

Stock removal tolerance = ±.25 rough cut
±.05 finish cut
±.30 total

Minimum stock removed = .10 given
Rough length = 25.0 finish length
0.1 minimum removed
0.3 mean of stock removal tolerance
25.4 ± .25

Example 2-1. Simple cut-off part

material being machined is predetermined. Here a minimum stock removal allowance of .10 is needed for a quality finish.

PREPARING A CHART

Example 2-2 uses a simple stepped shaft to illustrate the process chart format and the worksheet that supports its preparation. Although very experienced analysts can prepare a chart without the worksheet, it is recommended that the calculations be documented. This is useful when checking the work done, correcting errors or making changes at a later date.

A review of Example 2-2 shows that the chart is composed of three vertical sections.

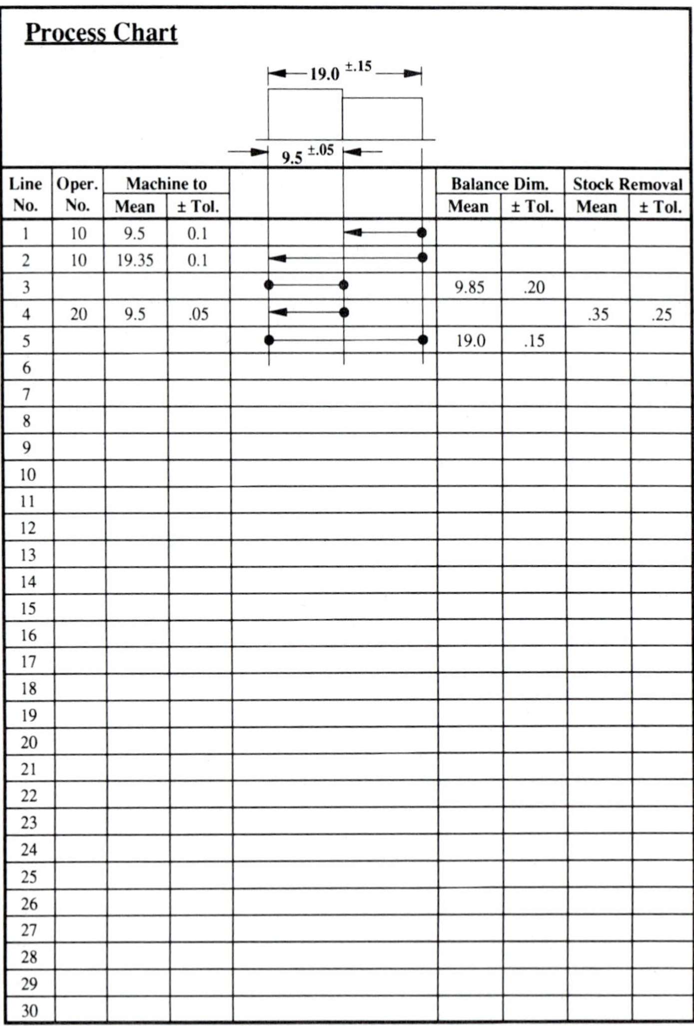

Process Chart

19.0 ±.15

9.5 ±.05

Line No.	Oper. No.	Machine to Mean	Machine to ± Tol.	Balance Dim. Mean	Balance Dim. ± Tol.	Stock Removal Mean	Stock Removal ± Tol.
1	10	9.5	0.1				
2	10	19.35	0.1				
3				9.85	.20		
4	20	9.5	.05			.35	.25
5				19.0	.15		
6							
7							
8							
9							
10							
11							
12							
13							
14							
15							
16							
17							
18							
19							
20							
21							
22							
23							
24							
25							
26							
27							
28							
29							
30							

Example 2-2.

Tolerance Assignment Worksheet							
To determine	**Line**	**Add/Sub/ Assign**	**Parameter**	**Line**	**From /To**	**Parameter**	**Line**
Balance dim.	5				From	Spec	
Balance tol.	5				From	Spec	
Machine dim.	4				From	Spec	
Machine tol.	4				From	Spec	
Machine tol.	1	Sub	Machine tol.	4	From	Balance tol.	5
Machine tol.	2	Assign	.10				
Balance tol.	3	Add	Machine tol.	1	To	Machine tol.	2
Removal tol.	4	Add	Balance tol.	3	To	Machine tol.	4
Removal mean	4	Add	.10 min.		To	Rem. tol.	4
Balance dim.	3	Add	Removal mean	4	To	Machine dim.	4
Machine dim.	1	Sub	Machine dim.	4	From	Balance dim.	5
Machine dim.	2	Add	Balance dim.	3	To	Machine dim.	1

Example 2-2 (cont.).

The center section graphically displays the individual operations by extension lines drawn for each step or shoulder of the part sketched at the top of the page. The left section lists machining dimensions and tolerances, while the right side lists balance dimensions and stock removal data. As the chart is developed, machine operations are noted by an arrow. A dot on the same horizontal line indicates the starting point, or datum, for the operation.

Balance dimensions are indicated by a horizontal line with a dot at each end. A balance dimension is the sum or difference of two or more machining dimensions. As such, the tolerance of the component dimensions always add to achieve the balance tolerance. Occasionally, a balance tolerance will be used as part of the stack-up to achieve another balance dimension. Balance dimensions can be developed when affected surfaces are defined, but a direct dimension is not available, or to calculate stock removal parameters.

The basic method of developing a process chart is given by the sequence below:

1. Fill in the process steps (machine operations) by marking the appropriate arrows.
2. Leave space to insert balance dimensions as the calculation is being completed.
3. Find the balance and machine dimensions that correspond to the drawing dimensions.
4. Compute tolerances first.
5. Determine stock removal values.
6. Determine the machine dimensions.

The part in Example 2-2 is machined in two steps. The first operation turns the hub (line 1), turns the O.D., and cuts off the part (line 2). The second operation recuts the end from the shoulder (line 4) refining the previous rough cut. Consequently, the drawing dimensions (labeled SPEC for specification) and tolerances are known and are entered on lines 4 and 5. Considering the tolerance first, line 1 is derived from lines 4 and 5. Line 2 is an assigned tolerance reflecting the capability of the machinery used. The balance tolerance, line 3, is the sum of lines 1 and 2. Adding the machine tolerance of line 4 results in the stock removal tolerance.

To develop the mean dimensions, an assigned minimum stock removal value of .10 is added to the stock removal tolerance to develop the stock removal mean. Adding this value to the mean dimension of line 4 results in balance dimension 3. Mean dimension 1 is based upon drawing dimensions, lines 4 and 5, while mean dimension 2 is based upon lines 1 and 3.

COMPLEX EXAMPLES

Examples 2-3 and 2-4 typify more complex process charts. The chart of Example 2-3a shows how the process is sketched out prior to any of the calculations. This part is machined in three operations. The first operation (10) roughs the part all over. The second operation (20) refinishes the left end. The final operation (30) completes the groove and the right end. A machine tolerance of .2 is assigned to operations 10 and 20.

Example 2-3b shows how dimensions and tolerances are entered. In this case a problem was encountered at lines 11 and 12. The zero balance tolerance at line 12 results in a negative tolerance at line 11. This indicates that the process chosen can not achieve the product tolerances. In this case, it was determined that the ± .2 tolerance on the 12.7 dimension could be increased to ± .4. The chart was changed accordingly and completed as shown in Example 2-3c. When such problems occur, they are resolved by changing product or machining tolerances, and/or the machining sequence. An area requiring close attention is the computation of the balance dimension on line 10. Here the stock removal mean is subtracted from the machining dimension to reflect the fact that the machining operation enlarges the groove.

The stepped shaft of Example 2-4 is machined in 3 operations, as shown by the grouped dimensions in the chart (operations 10, 20, 30). General machining tolerances of 0.1 were used for operation 10. In operation 20, it was observed that the tolerance on lines 8 and 9 must sum to the balance dimension on line 10, hence tolerances 8 and 9 were assigned the smaller value of 0.05.

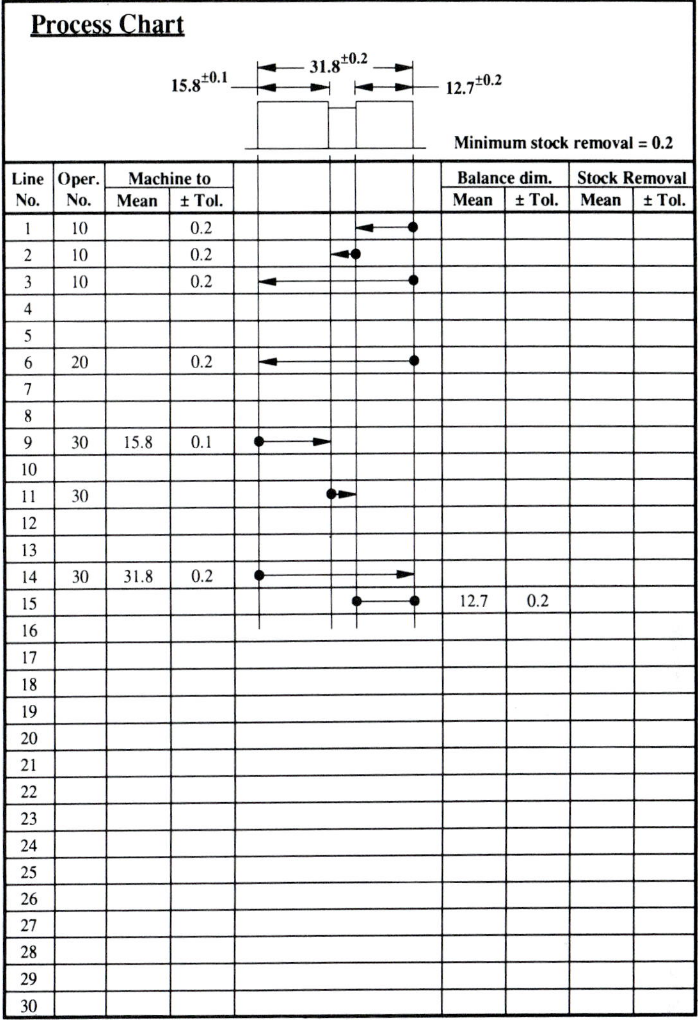

Process Chart

Line No.	Oper. No.	Machine to Mean	Machine to ± Tol.		Balance dim. Mean	Balance dim. ± Tol.	Stock Removal Mean	Stock Removal ± Tol.
1	10		0.2					
2	10		0.2					
3	10		0.2					
4								
5								
6	20		0.2					
7								
8								
9	30	15.8	0.1					
10								
11	30							
12								
13								
14	30	31.8	0.2					
15					12.7	0.2		
16								
17								
18								
19								
20								
21								
22								
23								
24								
25								
26								
27								
28								
29								
30								

Example 2-3

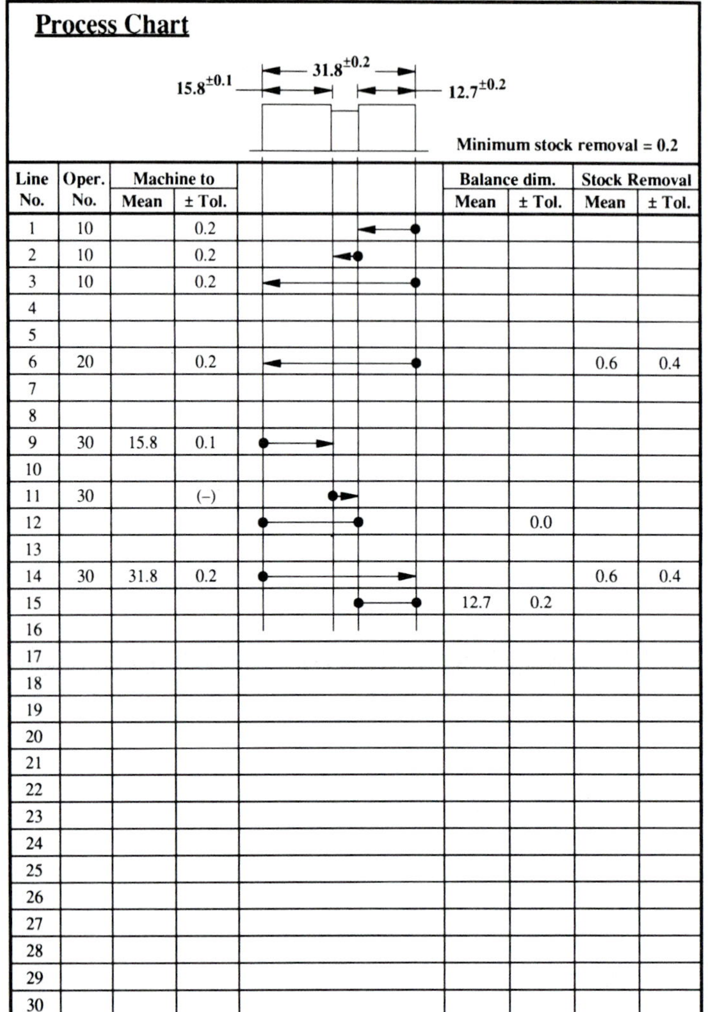

Process Chart

$15.8^{\pm 0.1}$ $31.8^{\pm 0.2}$ $12.7^{\pm 0.2}$

Minimum stock removal = 0.2

Line No.	Oper. No.	Machine to Mean	Machine to ± Tol.		Balance dim. Mean	Balance dim. ± Tol.	Stock Removal Mean	Stock Removal ± Tol.
1	10		0.2					
2	10		0.2					
3	10		0.2					
4								
5								
6	20		0.2				0.6	0.4
7								
8								
9	30	15.8	0.1					
10								
11	30		(–)					
12						0.0		
13								
14	30	31.8	0.2				0.6	0.4
15					12.7	0.2		
16								
17								
18								
19								
20								
21								
22								
23								
24								
25								
26								
27								
28								
29								
30								

Example 2-3a

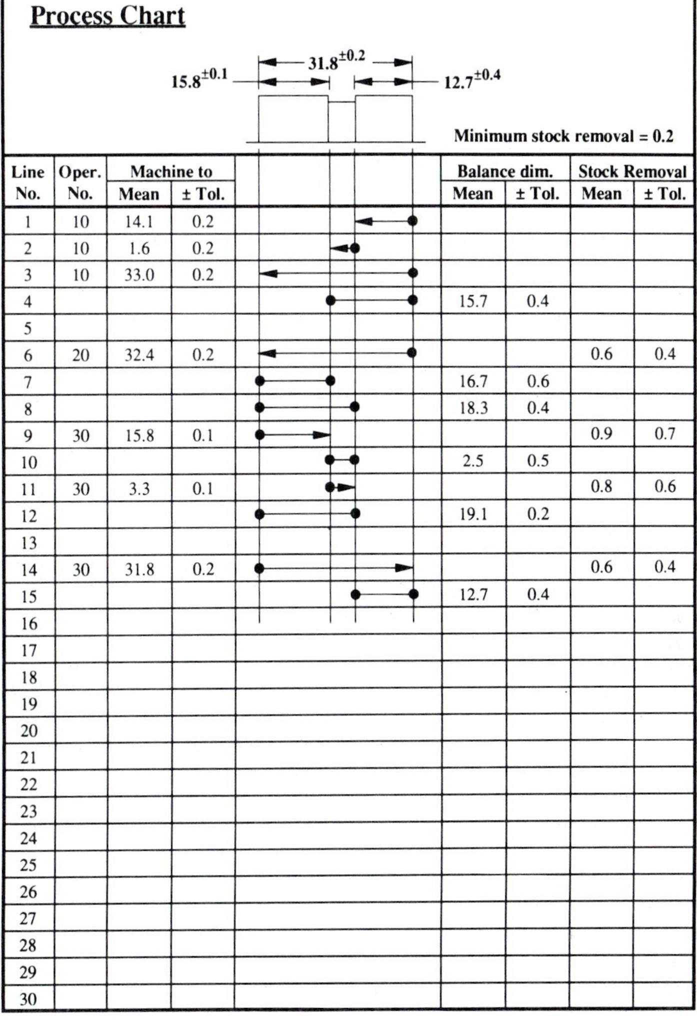

Line No.	Oper. No.	Machine to Mean	Machine to ± Tol.	Balance dim. Mean	Balance dim. ± Tol.	Stock Removal Mean	Stock Removal ± Tol.
1	10	14.1	0.2				
2	10	1.6	0.2				
3	10	33.0	0.2				
4				15.7	0.4		
5							
6	20	32.4	0.2			0.6	0.4
7				16.7	0.6		
8				18.3	0.4		
9	30	15.8	0.1			0.9	0.7
10				2.5	0.5		
11	30	3.3	0.1			0.8	0.6
12				19.1	0.2		
13							
14	30	31.8	0.2			0.6	0.4
15				12.7	0.4		
16							
17							
18							
19							
20							
21							
22							
23							
24							
25							
26							
27							
28							
29							
30							

Example 2-3b

Tolerance Assignment Worksheet							
To determine	**Line**	**Add/Sub/ Assign**	**Parameter**	**Line**	**From /To**	**Parameter**	**Line**
Machine dim.	14				From	Spec	
Machine tol.	14				From	Spec	
Balance dim.	15				From	Spec	
Balance tol.	15				From	Spec	
Machine dim.	9				From	Spec	
Machine tol.	9				From	Spec	
Stock rem. tol.	14	Add	Machine tol.	14	To	Machine tol.	6
Stk. rem. mean	14	Assign	0.2 min.				
Stock rem. tol.	6	Add	Machine tol.	6	To	Machine tol.	3
Stk. rem. mean	3	Assign	0.2 min.				
Balance tol.	12	Sub	Balance tol.	15	From	Machine tol.	14
Machine tol.	11	Sub	Machine tol.	9	From	Balance tol.	12
Balance tol.	4	Add	Machine tol.	1	To	Machine tol.	2
Balance tol.	7	Add	Machine tol.	6	To	Balance tol.	4
Stock rem. tol.	9	Add	Machine tol.	9	To	Balance tol	7
Stk. rem. mean	9	Assign	0.2 min.				
Balance tol.	8	Add	Machine tol.	1	To	Machine tol.	6
Balance tol.	10	Add	Balance tol.	8	To	Machine tol.	9
Stock rem. tol.	11	Add	Machine tol.	11	To	Balance tol.	10
Stk. rem. mean	11	Add	0.2 min.		To	Rem. tol.	11
Machine dim.	6	Add	Stock removal	14	To	Machine dim.	14
Machine dim.	3	Add	Stock removal	6	To	Machine dim.	6
Balance dim.	12	Sub	Balance dim.	15	From	Machine dim.	14
Machine dim.	11	Sub	Machine dim.	9	From	Balance dim.	12
Balance dim.	7	Add	Stock removal	9	To	Machine dim.	9
Balance dim.	4	Sub	Balance dim.	7	From	Machine dim.	6
Balance dim.	10	Sub	Stock removal	11	From	Machine dim.	11
Balance dim.	8	Add	Machine dim.	9	To	Balance dim.	10
Machine dim.	1	Sub	Balance dim.	8	From	Machine dim.	6
Machine dim.	2	Sub	Machine dim.	1	From	Balance dim.	4

Example 2-3c

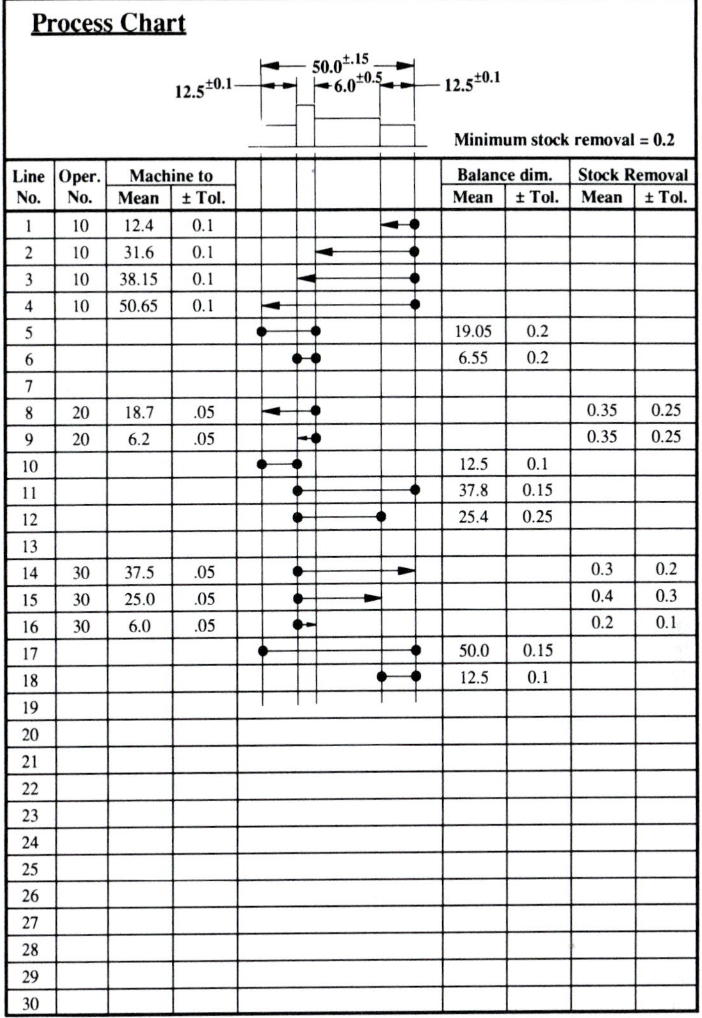

Line No.	Oper. No.	Machine to Mean	Machine to ± Tol.	Balance dim. Mean	Balance dim. ± Tol.	Stock Removal Mean	Stock Removal ± Tol.
1	10	12.4	0.1				
2	10	31.6	0.1				
3	10	38.15	0.1				
4	10	50.65	0.1				
5				19.05	0.2		
6				6.55	0.2		
7							
8	20	18.7	.05			0.35	0.25
9	20	6.2	.05			0.35	0.25
10				12.5	0.1		
11				37.8	0.15		
12				25.4	0.25		
13							
14	30	37.5	.05			0.3	0.2
15	30	25.0	.05			0.4	0.3
16	30	6.0	.05			0.2	0.1
17				50.0	0.15		
18				12.5	0.1		
19							
20							
21							
22							
23							
24							
25							
26							
27							
28							
29							
30							

Example 2-4

Tolerance Assignment Worksheet							
To determine	**Line**	**Add/Sub/ Assign**	**Parameter**	**Line**	**From /To**	**Parameter**	**Line**
Dim. & Tol.	18				From	Spec	
Dim. & Tol.	17				From	Spec	
Dim. & Tol.	16				From	Spec	
Dim. & Tol.	10				From	Spec	
Machine tol.	1-4	Assign	0.1				
Machine tol.	14	Sub	Balance tol.	10	From	Balance tol.	17
Machine tol.	15	Sub	Machine tol.	14	From	Balance tol.	18
Balance tol.	5	Add	Machine tol.	2	To	Machine tol.	4
Balance tol.	6	Add	Machine tol.	3	To	Machine tol.	2
Machine tol.	8	Assign	.05				
Machine tol.	9	Sub	Machine tol.	8	From	Balance tol.	10
Stock rem. tol.	8	Add	Balance tol.	5	To	Machine tol.	8
Stock rem. tol.	9	Add	Balance tol.	6	To	Machine tol.	9
Stk. rem. mean	8,9	Assign	0.1 min.				
Balance tol.	11	Add	Machine tol.	2	To	Machine tol.	9
Balance tol.	12	Add	Machine tol.	1	To	Balance tol.	11
Stock rem. tol.	16	Add	Machine tol.	9	To	Machine tol.	16
Stock rem. tol.	15	Add	Balance tol.	12	To	Machine tol.	15
Stock rem. tol.	14	Add	Balance tol.	11	To	Machine tol.	14
Stk. rem. mean	14-16	Add	0.10 min.		To	Rem. tol.	14-16
Machine dim.	14	Sub	Balance dim.	10	From	Balance dim.	17
Machine dim.	15	Sub	Balance dim.	18	From	Machine dim.	14
Balance dim.	11	Add	Stock rem. min.	14	To	Machine dim.	14
Balance dim.	12	Add	Stock rem. min.	15	To	Machine dim.	15
Machine dim.	9	Add	Stock rem. min.	16	To	Machine dim.	16
Machine dim.	8	Add	Machine dim.	9	To	Balance dim.	10
Balance dim.	5	Add	Stock rem. min.	8	To	Machine dim.	8
Balance dim.	6	Add	Stock rem. min.	9	To	Machine dim.	9
Machine dim.	2	Sub	Machine dim.	9	From	Balance dim.	11
Machine dim.	3	Add	Machine dim.	2	To	Balance dim.	6
Machine dim.	4	Add	Machine dim.	2	To	Balance dim.	5
Machine dim.	1	Sub	Balance dim.	12	From	Balance dim.	11

Example 2-4 (cont.).

SUMMARY

Process charting is a logical extension of the variation analysis methods of Chapter 1. As outlined in that chapter, bilateral tolerances are used to avoid errors possible in a complex analysis. Such tolerances are always additive as one progresses toward completion of the part. Mean dimensions are the sum or difference of two dimensional values, either machining or balance dimensions. Balance dimensions are usually placed ahead of a machine dimension so as to establish the stock removal allowance for the subsequent machining operation. As seen in the examples, the tolerances are computed before dimensions and the chart generally develops from the bottom up. Once complete, the chart can be checked by following the stack-up and dimensional changes from the top down.

Chapter 3

Statistical Analysis

The first two chapters dealt with analysis of accumulated tolerance based upon linear analysis techniques. In many instances, statistical methods are more appropriate for the analysis of complex variations. We will review a number of such techniques, following the definition of some basic statistical distributions. Our focus here is upon application of such techniques. The reader is referred to accepted works on statistical methods (of which there are many) for specific details as to the development of the underlying mathematics.

FREQUENCY DISTRIBUTION

If you make measurements of a parameter (such as a part dimension) over a production run of a sufficient number of pieces, you will find that it usually conforms to one of several familiar frequency distributions. A *frequency distribution*, such as the normal curve in Figure 3-1, can be plotted upon a histogram.[1] A *histogram* is developed by collecting all of the measurements that fall within certain bands (in this case, 1.0 divisions) and by plotting their frequency (number of occurrences) on the vertical axis. Statistical methods (and statistical computer programs) will fit an appropriate distribution to given histogram data. Other common frequency distributions are shown in Figure 3-2.

[1]Detailed information on histograms and frequency diagrams can be found in any introductory statistics text.

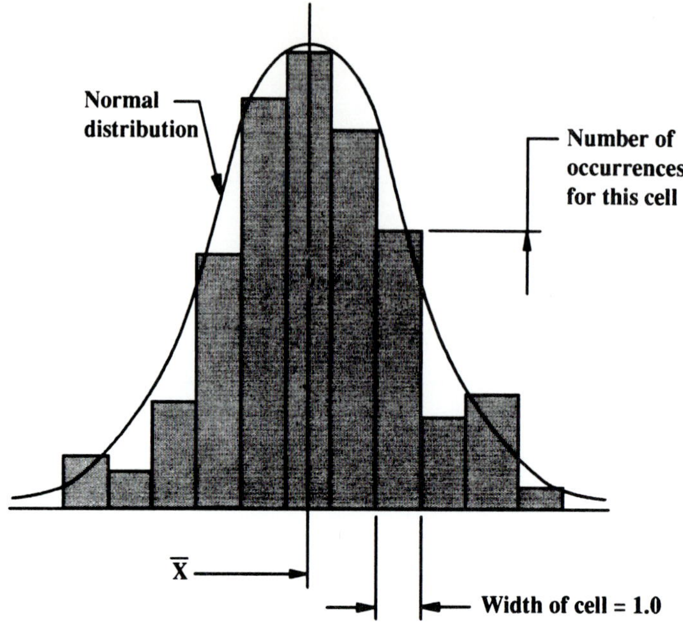

Figure 3-1. Normal frequency distribution fitted to histogram

As we contrast the various distributions, we note the following characteristics:

Normal. Most of the observed measurements occur at or near the mean ($\overline{X}$) with only a few occurrences at either extreme.

Rectangular. The observed measurements occur with equal frequency at any point within the curve.

Skewed. This distribution is not symmetrical like the others and the majority of the observations occur at one side or the other.

Normal distributions describe random events well. Operations such as placing a part at a target location tend to be random as there is no mechanism to bias the results in a particular direction. *Rectangular distributions* describe equal probability events. A tool that wears producing a dimension that changes uniformly during the sampling period appears to produce a rectangular distribution. However, such occurrences are time dependent and this seriously affects the assumption of statistical independence assumed in statistical analysis. *Skewed distributions* describe processes that are heavily biased to one extreme (e.g., a process that is monitored and adjusted on a frequent basis). A process that has no negative values (such as flatness) is frequently skewed. A normal distribution that is truncated on one side by sorting will approximate this type of distribution.

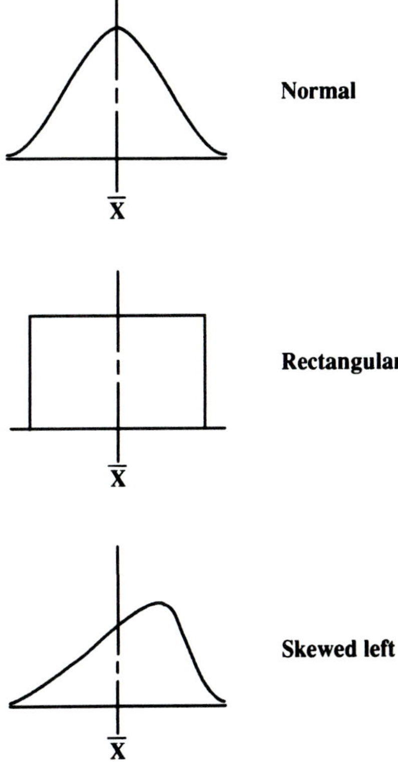

Figure 3-2. Common frequency distributions

STATISTICAL METHODS

When we do linear stack-ups, as in Chapter 1, we have made an underlying assumption that parts at extreme limits of tolerance will be assembled with other parts also at extreme limits. While this approach is conservative and can be effective in defining the worst case, it can be the source of new problems when process capabilities indicate large detail tolerance or when assembly requirements are exacting. If we would show this to a statistician, he would say that we are overly cautious and suggest a method frequently called RMS for *root mean square*, but more correctly called RSS for *root sum square*. This method assumes that we have parts made with dimensions whose variation has a normal distribution. (More on this key point follows in Chapter 4.) A simple example, shown in Example 3-1 serves to illustrate the difference between the linear and statistical methods. In linear analysis, we sum the mean dimensions to find the assembly limits. In the RSS method, the mean dimen-

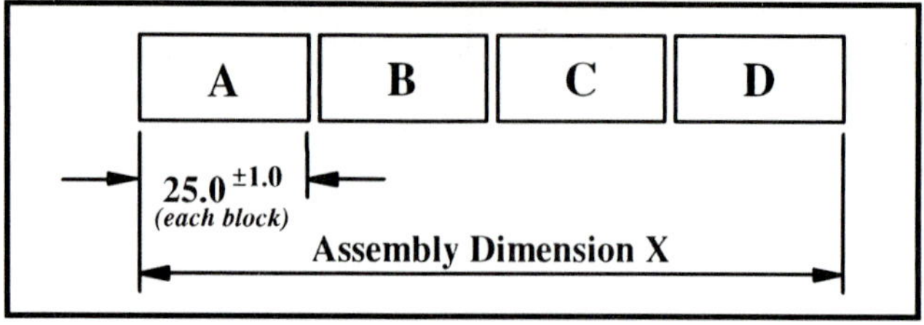

Linear Analysis:

$$X = \Sigma\, \text{mean} \pm \Sigma t_i$$
$$= 4\,(25) \pm 4\,(1.0)$$
$$= 100 \pm 4.0$$

Statistical Analysis:

Retain Component tolerance:

$$X = \Sigma\, \text{mean} \pm \sqrt{\Sigma t_i^2}$$
$$= 4\,(25) \pm \sqrt{4\,(1.0^2)}$$
$$= 100 \pm 2.0$$

Or, retain assembly limit:

let:

$$\sqrt{\Sigma t_i^2} = 4.0$$
$$\sqrt{4t^2} = 4.0$$
$$4t^2 = 16$$
$$t = 4.0$$
$$t = \pm\, 2.0$$

Example 3-1. Statistical analysis

sion is found in the same manner and has the same value. The statistically calculated tolerance is based upon the formula 3.1:[2]

$$T_{ASSY} = \sqrt{t_1^{\,2} + t_2^{\,2} + t_3^{\,2} + \ldots + t_n^{\,2}} \qquad (3\text{-}1)$$

where t_i = individual tolerances.

As seen in this example, the total variation of an assembly based upon statistical methods is significantly less than the linear method (only half for four equal tolerances). This provides two opportunities:

1. Retain the assigned component tolerances and reduce the assembly variation. This is important in stack-ups with lots of parts (e.g., in a transmission assembly with many

[2] Subject to the limitations discussed in Chapter 4.

metal and friction material clutch plates sandwiched together). Here, the problem alternates between having enough space for a thick clutch plate stack or having enough movement in the engagement mechanism to clamp them in place with worn, thin plates.

2. Retain the linear (or other design target) assembly limits and calculate larger component tolerances. Example 3-1 also illustrates this option and shows how the component tolerance (in this case) is doubled. This approach is almost always required when complex mechanisms with tight operational limits are toleranced, as limit analyses can produce unrealistically small component tolerances.

Example 3-2 is based upon Example 1-1. Here we see a real world type problem where the tolerances assumed to achieve control by limit analysis methods cannot be met in the manufacturing process. Statistical analysis is now used to compute the assembly requirements. This example also illustrates how real problems are combinations of linear and statistical analysis methods and how the analysis may require several iterations of assigning tolerance, completing the analysis, and making adjustments until a satisfactory balance of component tolerance and assembly requirements is met. One is cautioned that the statistical method is not employed so that the analysis would have more acceptable results, but rather a simple (linear analysis) method was used until it could not yield acceptable results. Then a more complex, presumably realistic method was chosen. As previously stated, compliance to the assumptions upon which statistical analysis is based is necessary for the results to be valid.

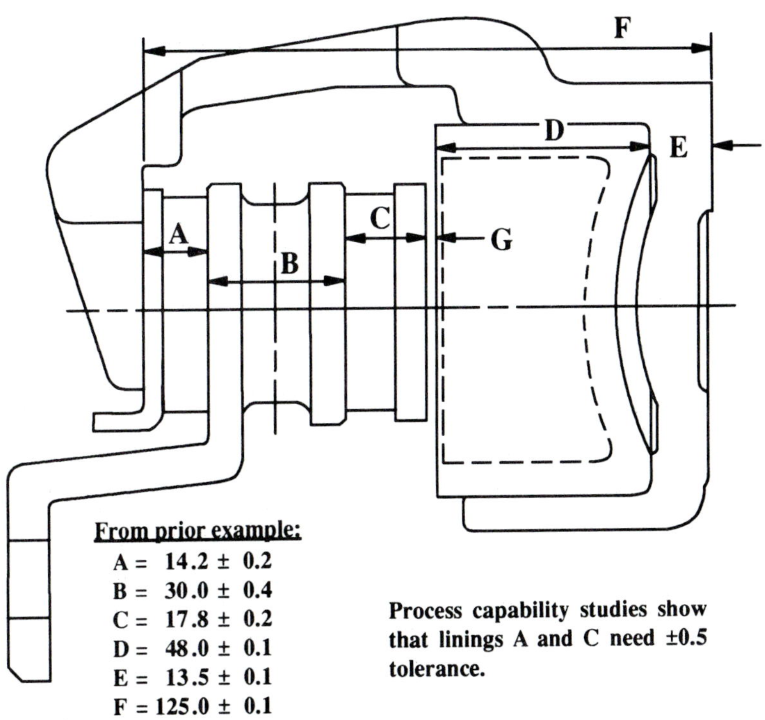

Example 3-2. Statistical analysis of disc brake

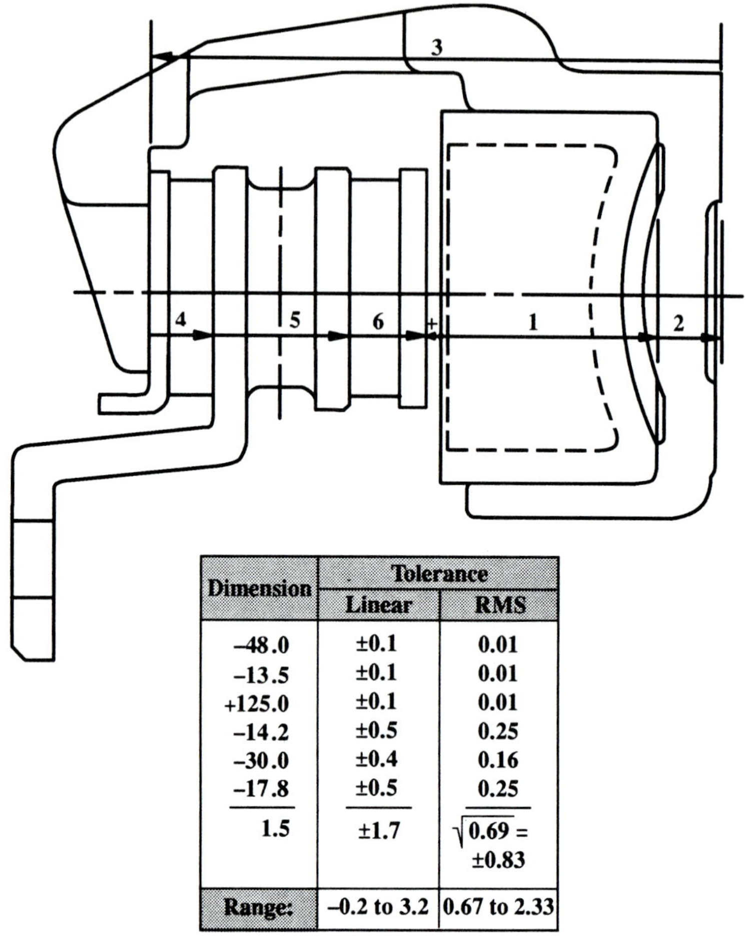

Dimension	Tolerance	
	Linear	RMS
–48.0	±0.1	0.01
–13.5	±0.1	0.01
+125.0	±0.1	0.01
–14.2	±0.5	0.25
–30.0	±0.4	0.16
–17.8	±0.5	0.25
1.5	±1.7	$\sqrt{0.69}$ = ±0.83
Range:	–0.2 to 3.2	0.67 to 2.33

Example 3-2. Solution

BENDER AND GILSON METHODS

In many cases, the manufacturing controls that will be outlined in Chapter 4 are not in place. Conversely, an engineer may be analyzing an assembly, but have no input from the manufacturing engineers involved. Must he/she then revert to a full linear stackup? No say Messrs. Bender, Gilson, and a host of other engineers with a practical viewpoint.[3] Based upon the rational of Figure 3-5 and practical experience, Bender proposes that:

$$T_{ASSAY} = 1.5\sqrt{\Sigma t_i^{\,2}} \qquad (3\text{-}2)$$

Figure 3-3 shows the extreme variations between linear and statistical accumulations

[3]Bender, Jr., Arthur, "Benderizing Tolerances," *Graphic Science,* December, 1962.

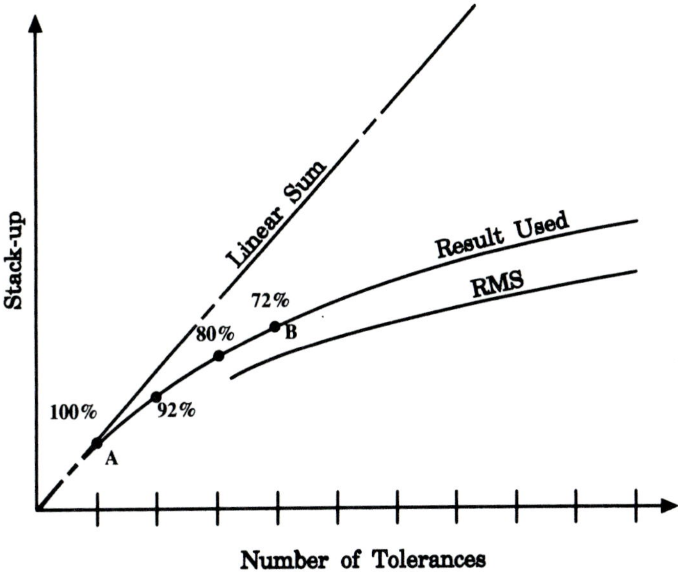

Result A **is the full linear stack-up**
Result A → B is the linear stack-up times Gilson factor
Result B → is a Benderized (x 1.5) RMS value

Figure 3-3. Middle-of-the road statistics

as the number of components increase. Finding the linear result too large and the RSS result too small, a modified middle-of-the-road answer seems appropriate. This method works well and others have developed their own factor to replace the 1.5 "Benderizer" based upon observation of their own processes. You should note that this approach doesn't work for small numbers of components, sometimes producing a number larger than the linear value. In that case, the linear value should be used.

J. Gilson[4] arrives at a method which produces similar results. However, he applies his factor to the linear stack-up. His factors are summarized in Figure 3-4.

This author has used the two methods combined to cover the area where the Bender Method does not work well and to test for stack-up wherein a few tolerances dominate the result and hence affect the independence assumption for the assembly. The following test is used to decide between the two methods.

Test

1. Calculate 85% of the sum of the RSS column.
2. Rank the largest 4 entries in the RSS column beginning with the largest value. If one part is used several times, enter it *only once* as the tolerances may be dependent.

[4]Gilson, J. *A New Approach to Engineering Tolerances,* The Machinery Publishing Co. Ltd., London, 1951.

Tolerance Accumulation on Assemblies

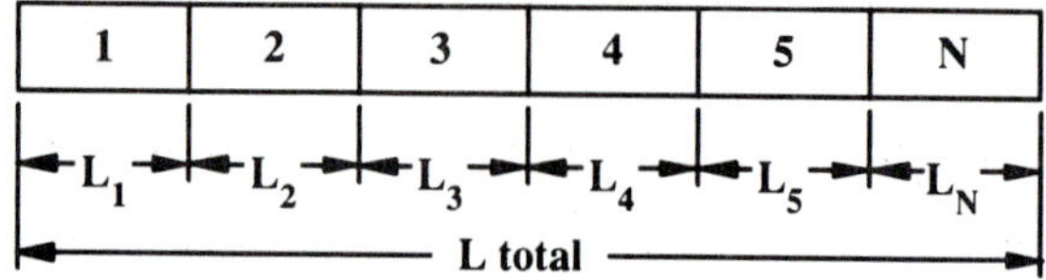

"T" tolerance variation for (L total) length of "N" assembled items of dimensions L_1, L_2, L_3....L_N and tolerances t_1, t_2, t_3....t_N.

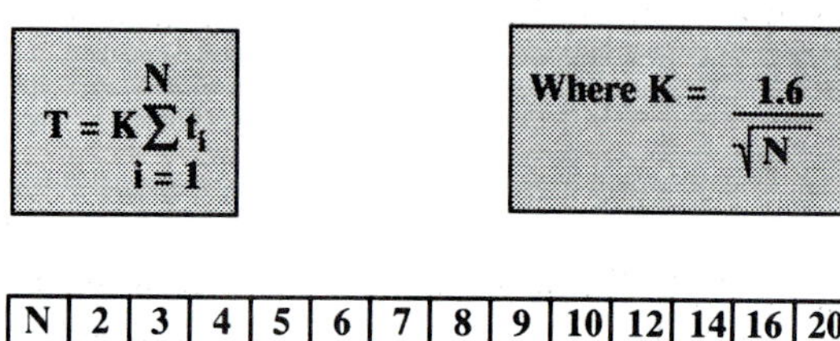

N	2	3	4	5	6	7	8	9	10	12	14	16	20
K	*	.92	.80	.72	.65	.60	.56	.53	.50	.46	.43	.40	.36

*** Use standard stack-up methods (K = 1.0)**

Figure 3-4. Gilson Factors

3. Add one value at a time until the 85% of the RSS sum is exceeded. The number of values added = N.
   a. If N is 1, 2, 3, or 4, find K

N	1	2	3	4
K	1.0	.92	.80	.72

   Result = K times linear sum.

   b. If N = 5 or more:

   Result = 1.5 times RSS sum.

Example 3-3 illustrates this approach. For large N values ($N > 4$) a Benderized RSS result is used. This value is the expected statistical stack-up increased by 50% to be more conservative due to many possible factors which can cause the statistical result to be overly optimistic. For $N = 1$ the full range linear stack-up result is used. Between these two ranges, a compromise result is obtained by multiplying the full linear stack-up by a Gilson factor (K). (See Example 3-4.)

An additional purpose of the test is to determine if only a few tolerances dominate the result. Hence, it is possible to have an analysis with many entries (e.g., 20) and a low N value such as 2 or 3.

Example 3-5 illustrates how a lack of component conformance to assumed statistical

RSS CALCULATIONS

SUBJECT:	ORIGINAL	REVISED	
REAR WINDOW BLACK OUTLINE TO	DATE:		APPENDIX:
SIGHT LINE	ANALYST:		PAGE: 2

NO	DESCRIPTION	CF	± TOLERANCE LINEAR	± TOLERANCE SQUARED	MEAN SHIFT
	(CONT.)				
10.	LOCATION OF BLACKOUT LINE ON GLASS (1.5 SIN 29°)	3	0.7272	0.5288	
11.				0.2461	
12.				0.2500	
13.				0.2500	
14.	THICKNESS OF SPACE	3	0.5000	0.2500	
15.	FORM TOLERANCE GLASS (2 SIN 61°)	3	1.7492	3.0597	
16.	LOCATION CHIMSEL MTG SURFACE (5 SIN 85°)	4	0.4981	0.2481	
16.	LOCATION OF SIGHT LINE ON CHIMSEL (1.5 SIN 85°)	4	1.4943	2.2329	
	SUM		7.8849	7.3331	-0.7272
	RSS √			2.7074	
	BENDERIZER/GILSON FACTOR X 1.5			4.0611	
	VARIATION +/-		4.1		

0.85 x 7.3331 = 6.2331

3.0597 + 2.2329 + 0.5288 + 0.2461 < 6.2331

Use Benderizer Factor

Example 3-3. Benderizing Factor

distributions affects a stack-up. This example is based upon Example 3-1, which indicates that 4 components with a 25mm mean and tolerance limits of 1 had an expected assembly size of 100 ± 2. Although the components in Example 3-5 meet detail requirements, 50% of the assemblies exceed the expected upper limit of 102. All of the assemblies, however, fall within the 103 limit established by Benderizing the ± 2 RSS tolerance.

In this example, the assembly results deviated from those expected because the component distributions were not properly centered at their design mean value. The effect of improper targeting is the most prevalent problem when statistical analysis methods are used. This can result, as in the example, when process capability is better than the allowed tolerance and the controls used focus on being *within tolerance* versus *on target*. Further

RSS CALCULATIONS

SUBJECT:	ORIGINAL	REVISED	
UP/DOWN VARIATION ROOF INNER	DATE:		APPENDIX:
TO OUTER	ANALYST:		PAGE:

NO	DESCRIPTION	C F	± TOLERANCE LINEAR	± TOLERANCE SQUARED	MEAN SHIFT
1.	LOCATION TOLERANCE ROOF OUTER	1	0.50	0.25	
2.	FORM TOLERANCE ROOF INNER	2	0.50	0.25	
3.	LOCATION TOLERANCE ROOF INNER	3	1.00	1.00	

0.85 x 1.50 = 1.275

1.00 + 0.25 + 0.25 > 1.275

N = 3 K = 0.80

	LINEAR	SQUARED	MEAN SHIFT
SUM	2.00	1.50	
RSS √			
BENDERIZER/GILSON FACTOR X 0.80	1.60		
VARIATION +/-	1.60		

Example 3-4. Gilson Factor

examination of the effect of improper targeting shows that the empirically derived Benderizing factor is, in fact, an expansion factor which compensates for this effect. Assuming components with equal tolerances, we can show that the 1.5 factor will set safe limits for off-target means ranging from 4 components at 50% of tolerance (this example) to 16 components at 84% of tolerance (capability indexes of $C_p = 1.20$ and $C_{pk} = 1.0$). When targeting error is controlled, as discussed in Chapter 4, it becomes possible to calculate the mean shifts and expansion factors are no longer required.

Absent other controls, the possibility of off-target components makes a good case for assembly inspection operations for statistically derived assembly tolerances, even in cases where the components have been inspected.

Consider the following component performance:

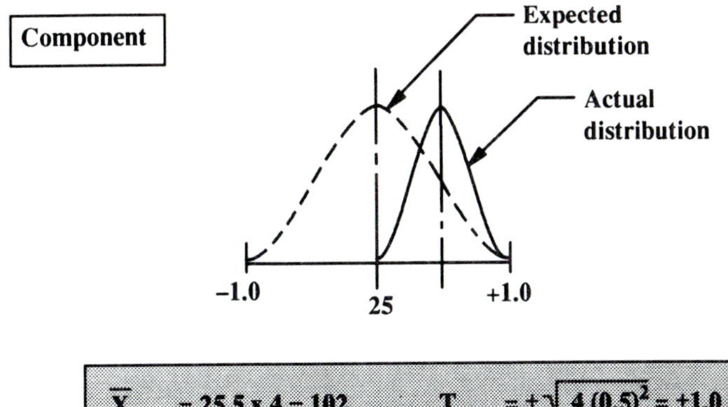

$\overline{X}_{ASM} = 25.5 \times 4 = 102 \qquad T_{ASM} = \pm\sqrt{4\,(0.5)^2} = \pm 1.0$

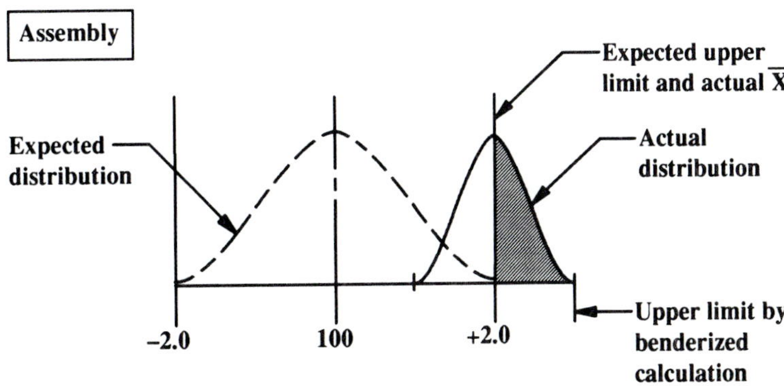

Example 3-5. Assembly mean at expected upper limit

Expected Variation

Further consideration of the contributing factors (tolerances) is called for when assembly dimensions are the result of both component tolerances and fixturing in the manufacturing process, as is the case with complex automotive body weldments. When a single fixture is used, its errors (tolerance) which are fixed do not induce a variable effect in the parts produced. However, these fixture variations do effect the mean value of the part variability within the probable assembly limits as in Figure 3-5. When the fixture tolerances are a significant contributor to the probable assembly dimension, the expected variation can be calculated by repeating the assembly dimension calculation and deleting or reducing the fixture tolerances used. Some analysts retain about 30% of the fixture tolerance to represent fixture repeatability. Figure 3-6 shows an example of this technique applied to an automotive door seal.

Chapter 4 defines *process capability* (C_p) and *process centering* (C_{pk}) indexes. Ex-

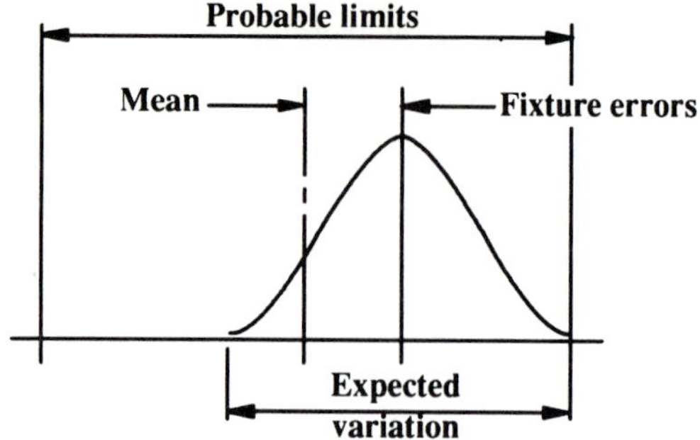

Figure 3-5. Expected variation vs. probable limits

pected variability is similar to the C_p value and the probable limits are similar to the C_{pk} value.

Contributing Factors

Identification of the major contributors to the assembly tolerance allows you to focus on those contributors for possible refinement during an analysis or to control the appropriate dimensions during the manufacturing phase. There are two recognized methods of determining *contributing factors*—linear and squared component ratios. In the first case, a percentage is calculated by dividing the component tolerance by the sum of the linear tolerances. The squared component tolerance method is similar but uses the squared component tolerance divided by the sum of the squared tolerances to compute the percentage contribution.

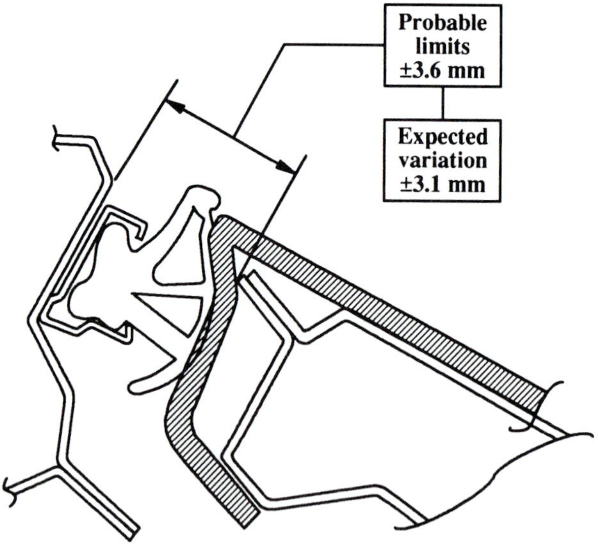

Figure 3-6. Expected variation notation

Since the contributing factors are normally used to rank tolerances, accuracy is not a major issue and either method is generally acceptable. The squared method, however, is preferred as it can be shown to be more consistent with statistical tolerance methods.

Scrutiny of the forms used in Examples 3-3 and 3-4 shows that they have a column to identify contributing factors (CF). Some of the available tolerance programs calculate both contributing factors and sensitivity factors. Sensitivity factors are computed by holding all tolerances except one at a fixed value and varying the tolerance under consideration by a fixed amount. The resulting assembly tolerances can then be compared to see if one tolerance has a more significant effect than another. Sensitivity effects usually occur in mechanisms where geometry effects are magnified.

Mixed Data

On occasion, a stack-up is based upon a mixture of measured statistical parameters and individual expected variations based upon tolerance values. This requires special handling when modifiers such as the Benderizer method is used. When this situation occurs, the stack-up result is derived by summing the squared values of the individual expected variations, multiplying this sum by 2.25 (1.5 Benderizer factor squared), and then adding the squared values of the measured parameters. The square root of the total sum is the modified RSS value. Additional Bender or Gilson factors are not applied. The mean value of the assembly must be based upon the actual means of the components where data is used.

SELECTIVE ASSEMBLY

Selective assembly is used in instances where tighter assembly tolerance control than available processes will yield is required. Figure 3-7 shows a hydraulic piston which requires a clearance (Figure 3-7a) of .013 to .030. Even though a good statistical distribution is known for the parts (Figure 3-7b) the maximum clearance resulting from setting the minimum clearance in a conventional manner is unacceptable. Only a few parts fall within the .030 limits indicating an assembly process that would not be feasible (see Figure 3-7c).

Selective assembly techniques are applied in Figure 3-8a. The distribution are overlapped and divided into sections (in this case, 4 sections of 1.5 σ). The divisions must be based upon σ of each distribution so as to yield the same number of parts for related sections. In this case, 7% each from a1-b1 and a4-b4, and 43% from a2-b2 and a3-b3. In practice, both parts are dimensionally sorted into 4 groups and assembled by group 1 $\rightarrow$ 1, 2 $\rightarrow$ 2, and so on. The percentage of parts within defined limits is readily found in a Z table, available in any statistical text book and included in the Appendix.

To make such systems work, the distributions must have equal or nearly equal widths. When the widths are not equal, the sections will produce different results when combined. The distributions in Figure 3-8a are slightly different in width. Figure 3-8b shows the assembly variation for different sections. If the width of the distributions is significantly different, they must be adjusted. Perhaps multiple runs at different mean settings made for one component as in Figure 3-9 is the answer.

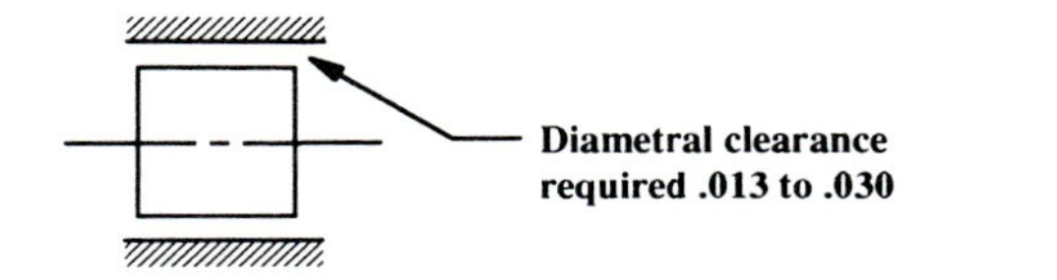

a) Hydraulic piston requirement

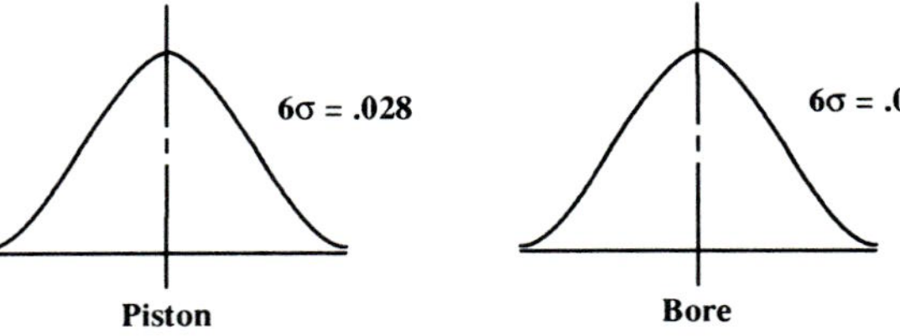

b) Process capability

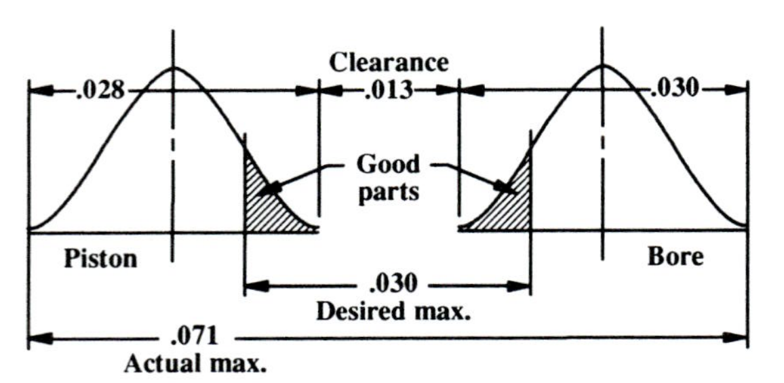

c) Assembly based upon .013 min.

Figure 3-7. Conventional assembly

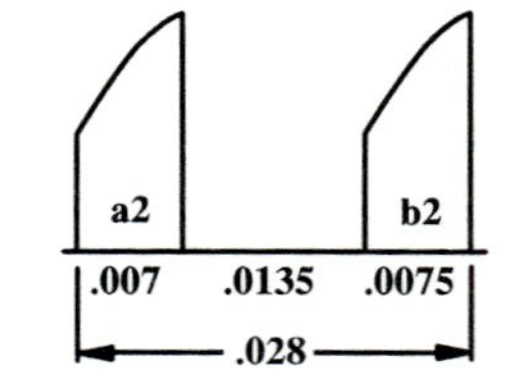

a) Overlapped distributions

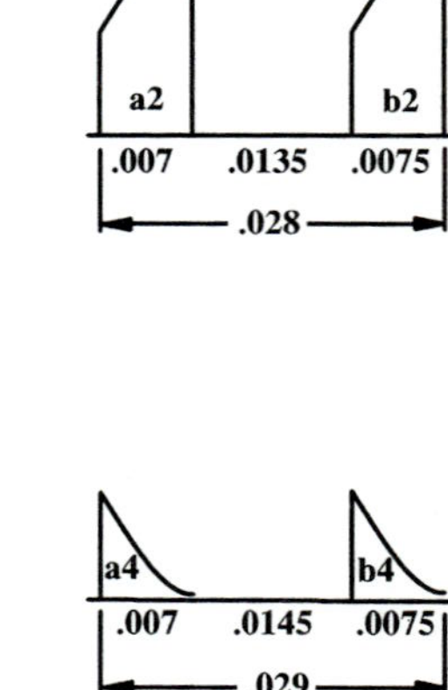

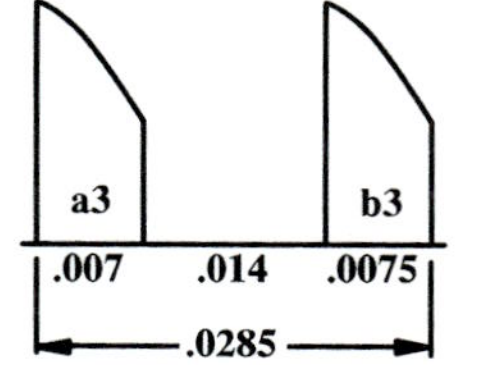

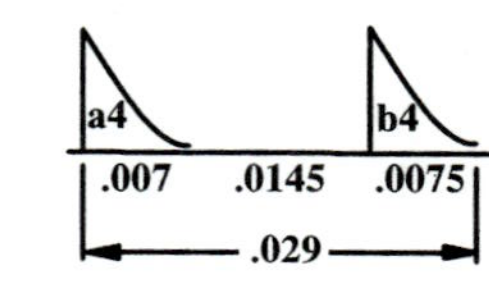

b) Paired sections

Figure 3-8. Selective assembly

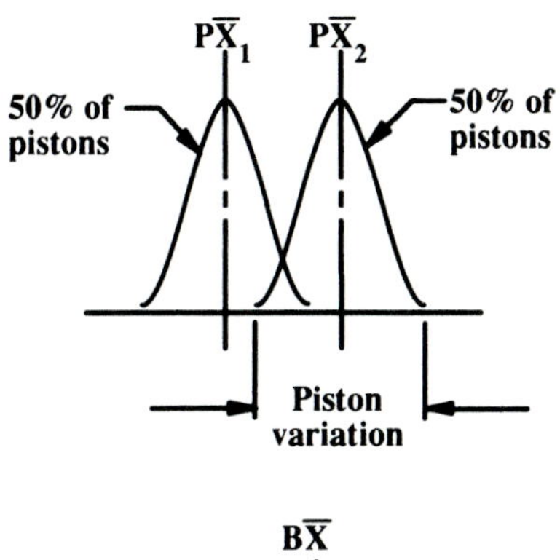

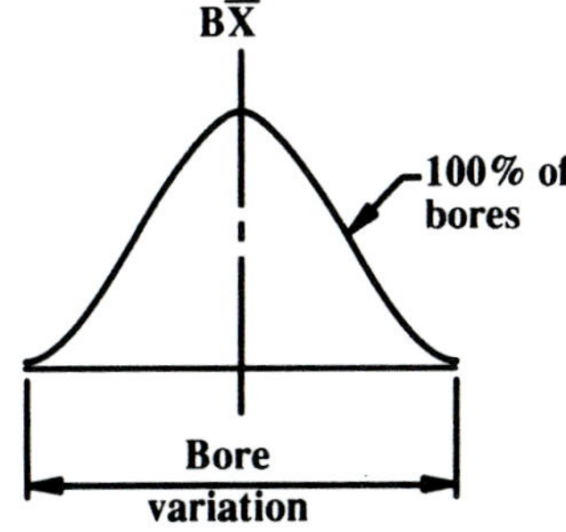

Figure 3-9. Selective assembly: adjusted means

SHIMMED ASSEMBLIES

Shimming is another technique which reduces tolerance accumulation to more acceptable limits. Development of such a method requires that the expected tolerance range be known; calculated by one of the techniques previously discussed. Also the acceptable tolerance must be known.

Figure 3-10a shows the stack-up for a bearing assembly[5] where the expected axial tolerance was 0.72mm based upon a probability calculation. The bearing required that end-play exist (*end-play* means assembly clearance, while interference is called *preload*). As will be seen, it is necessary to add shims when the assembly result falls in the preload region in order to obtain the required end-play. Dividing the 0.72 value by 3 produced a 0.24 end-play range which was found to be reasonable when compared to the bearing manufacturer's specifications. Adjustment of the mean dimensions of the assembly resulted in an assembly range of +0.24 to –0.48 as seen in Figure 3-10a. Reference to a Z table shows that 16% of the assemblies will lie in the acceptable end-play area and turn freely, while the remaining 84% will have a preload condition. When one 0.24 shim is added to the assemblies with preload, the mean shifts to the right and the center 68% will now have from 0 to 0.24 end-play and will turn freely. The remaining 16% can be moved into the acceptable end-play range by adding a second shim.

[5]This example is based upon *Engineering Bulletin No. 6* (Rev. A). Reprinted with the permission of The Timken Company, Canton, Ohio.

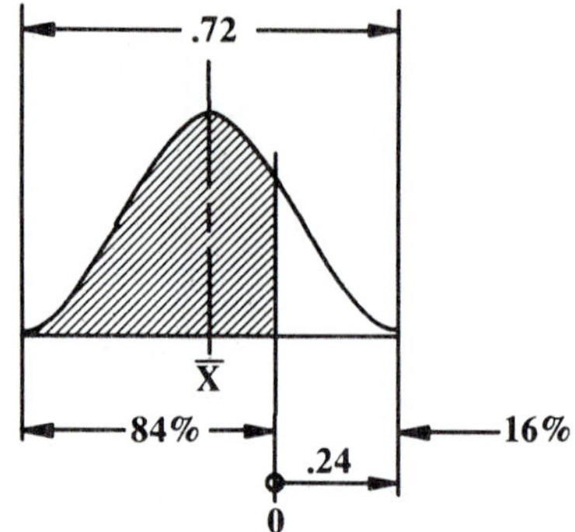

a) Bearing installation- no shims

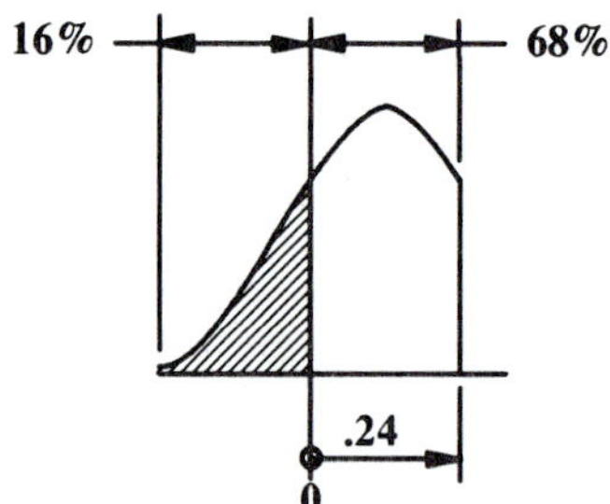

b) Bearing installation- one shim

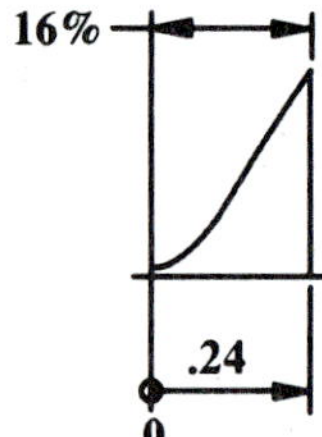

c) Bearing installation- two shims

Figure 3-10. Bearing end play

MONTE CARLO SIMULATION

When tolerance distributions are known (or assumed based upon similar processes), a *Monte Carlo simulation* can provide an accurate assessment of cumulative effects. The principle of this technique is to simulate sampling of known process distributions. The concept dates to the early 1900s, however, it has become a practical tool for variation management only recently because the technique, although conceptually simple, requires hundreds or thousands of calculations. Use of the technique expanded in the 1960s and 1970s on mainframe computers. Most recently, the expanded power of personal computers has made the technique usable by practically anyone.[6]

Underlying Theory

Using a two-component system as an example, we seek the solution to:

$$P(A + B = A_i + B_i) = PA_i \times PB_i \tag{3-3}$$

This reads: the probability that the sum of $A + B$ is equal to the sum of $A_i + B_i$ equals the probability of A_i times the probability of B_i. If we have known or assumed distributions for A and B, we can use the computer to select random numbers for i and sample the two distributions. Since sampling a frequency plot would typically yield two values for any given random number, cumulative distributions are used instead (see Figure 3-11).

The computer now performs the following tasks:

1. Determines PA_1 from random number and sample the distribution for A_1.
2. Determines PB_1 from another random number and sample the distribution for B_1.
3. Adds $A_1 + B_1$ = Value 1.
4. Multiplies $PA_1 \times PB_1$ = Probability 1.
5. Repeats hundreds or thousands of times.
6. Sums the probability for all values that are contained in a stated cell width.
7. Draws a histogram of the results.

Example 3-6 is a simulation of the probability calculation of Example 3-1. As can be seen, the results of the RSS and MCS study are essentially the same. This simple example does not do the MCS approach justice. When distributions are not normal and the problem complex, MCS yields results unattainable by RSS methods.

Example 3-7 studies the selective assembly problem of Figure 3-8a by use of MCS techniques. Following the example, as the maximum assembled clearance of 0.067 was judged excessive, the piston was manufactured at two mean values as in Figure 3-9. The assembly was partitioned into 4 groups as shown. Two sets now have maximums of 0.031, while the remaining two have maximums of 0.036. From the distributions developed by the MCS program, you can see that approximately 10% of set 1 and 5% of set 4 (or 5% of the total) exceed the desired 0.030 maximum. Study of the distributions suggest that further optimization is possible and that the desired fit range may be closely approximated.

Example 3-8 applies MCS modeling to a mechanism. Here the technique is extremely valuable as it permits the mechanism's tolerance effects to be studied at several points in its operating range, as well as permitting the use of different distributions for appropriate components.

[6]See Appendix for various available programs.

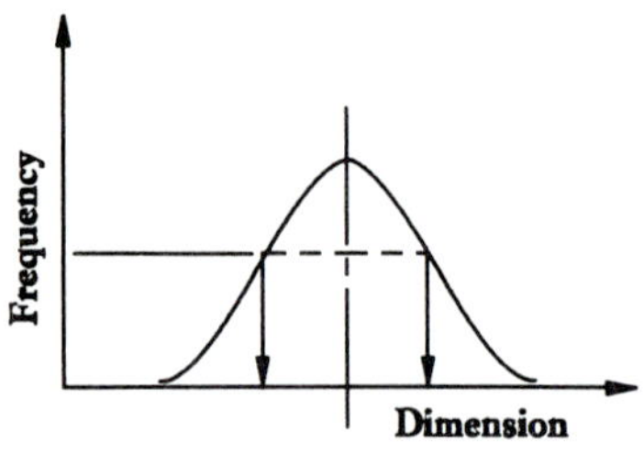

a) Frequency distributions do not yield unique values for a sample point

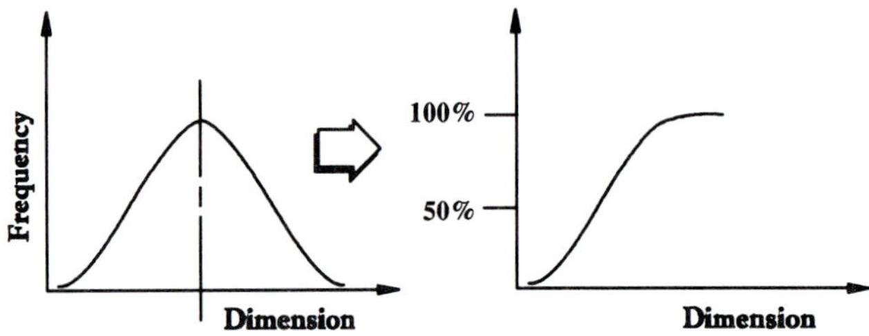

b) Convert from frequency to cumulative distribution

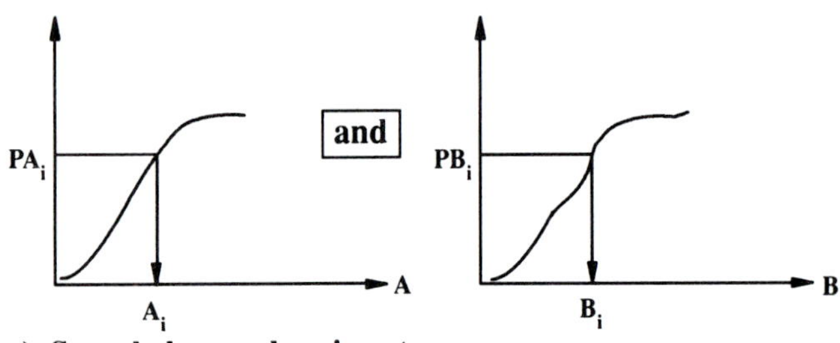

c) Sample by random input

Figure 3-11. Cumulative frequency distribution for MCS

MULTI-DIMENSIONAL PROBLEMS

The techniques and examples discussed thus far have related essentially to one-dimensional problems. In the real world, two- and three-dimensional problems are encountered sometimes. Such problems are most easily conceptualized by using simulation methods. In such cases, the equation defining the tolerance relation being evaluated simply reflects the contribution of the two- or three-dimensional aspects and the simulation will generate an appropriate answer.

```
SAMPLE SIZE     1000                                    SIMULATED    STATISTICAL

NOMINAL             100.0000  LOW                         97.5129      97.9722
MEAN                100.0166  HIGH                       102.0234     102.0610
STD DEV               0.6814  RANGE                        4.5105       4.0888
SKEWNESS             -0.0248  CAPABILITY INDEX
KURTOSIS             -0.0163  SHIFT FROM NOM               0.0166

DISTRIBUTION TYPE:
NORMAL (ASSUMED)
                              HISTOGRAM      1'*' = 6 SAMPLES (SIMULATED DATA
                                             '.' = BEST FIT DISTRIBUTION
CUM
PROB        MIDPOINT          FREQUENCY

0.000        97.1000            0 +
0.000        97.5000            1 +
0.001        97.9000            1 +
0.006        98.3000           12 +*.
0.027        98.7000           39 +*****.*
0.089        99.1000           82 +***************.
0.224        99.5000          177 +*****************************.*
0.432        99.9000          241 +**************************************.**
0.661       100.3000          198 +*********************************
0.852       100.7000          151 +************************.*
0.944       101.1000           69 +***********.*
0.985       101.5000           23 +***.
0.997       101.9000            6 +.
1.000       102.3000            0 +
1.000       102.7000            0 +
```

Example 3-6.

RSS methods can also be applied to this class of problem, however, the complexity increases. As with a simulation model, the equation defining the relationships must also be generated. Assuming that the equation is to be based upon variables X, Y, and Z, all with normal distributions, then:

$$T_{ASSAY} = \sqrt{\left(\frac{\delta}{\delta_x} T_x\right)^2 + \left(\frac{\delta}{\delta_y} T_y\right)^2 + \left(\frac{\delta}{\delta_z} T_z\right)^2} \tag{3-4}$$

where: $\frac{\delta}{\delta_x}$ is the partial derivative of the tolerance equation with respect to X, etc.

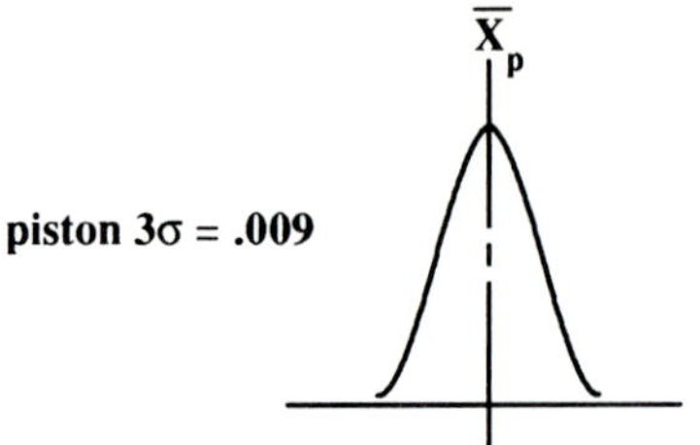

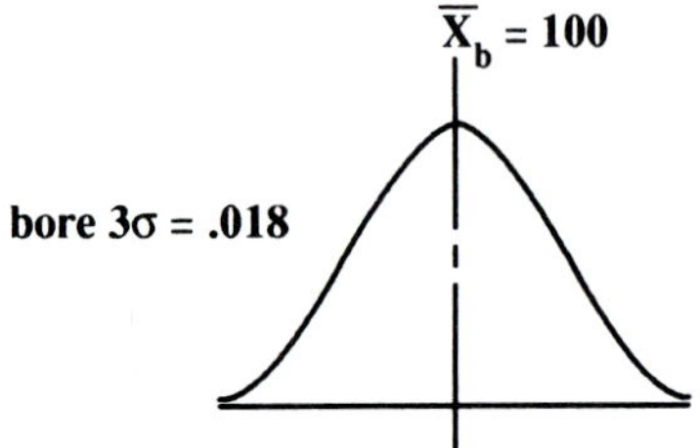

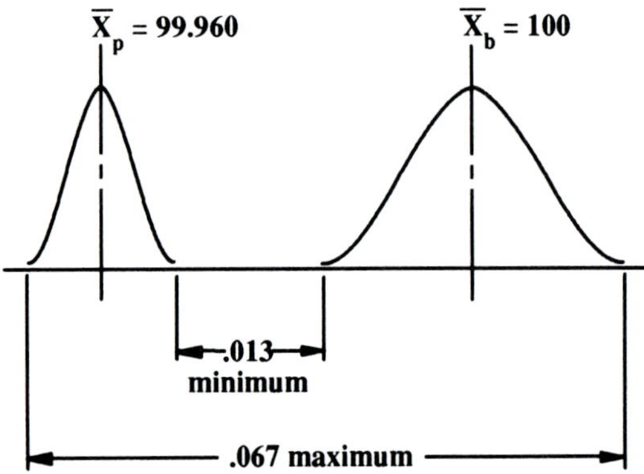

Example 3-7. Expected results of conventional assembly

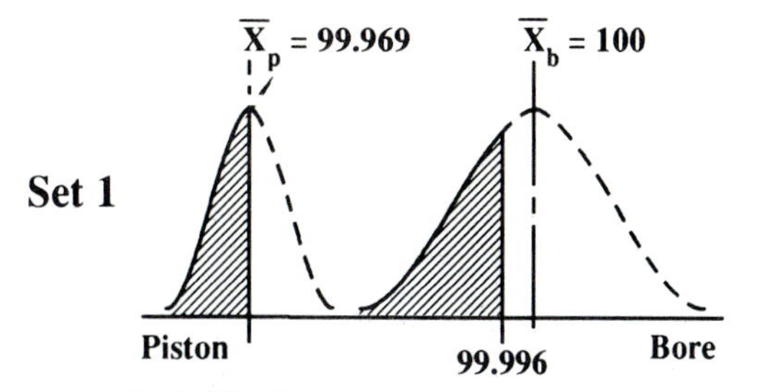

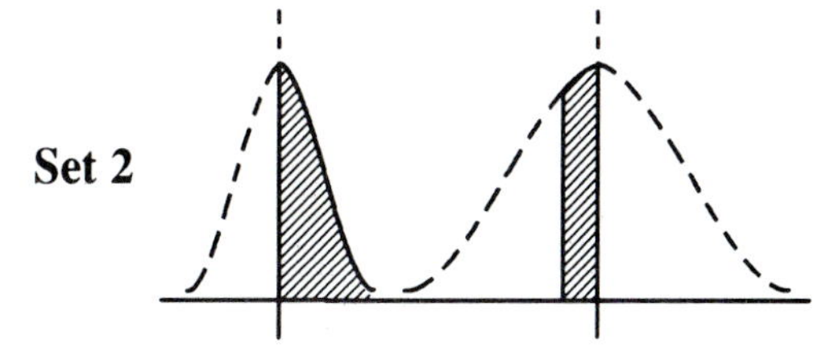

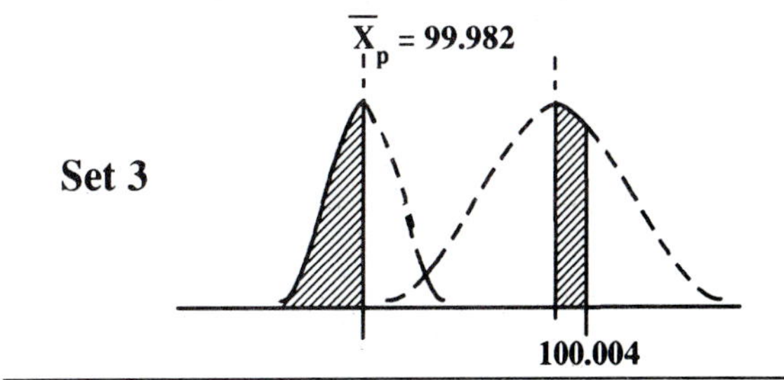

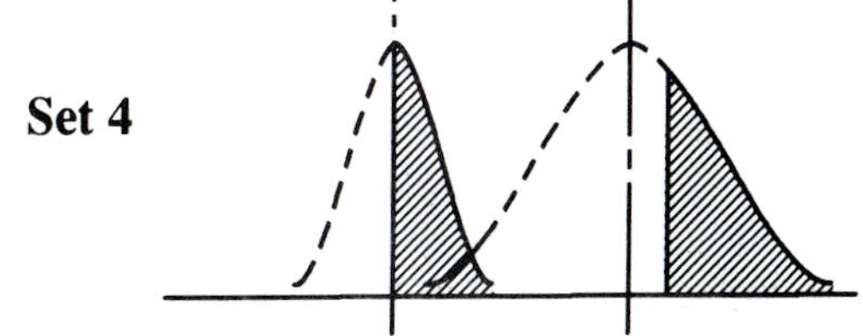

Example 3-7. (cont.)

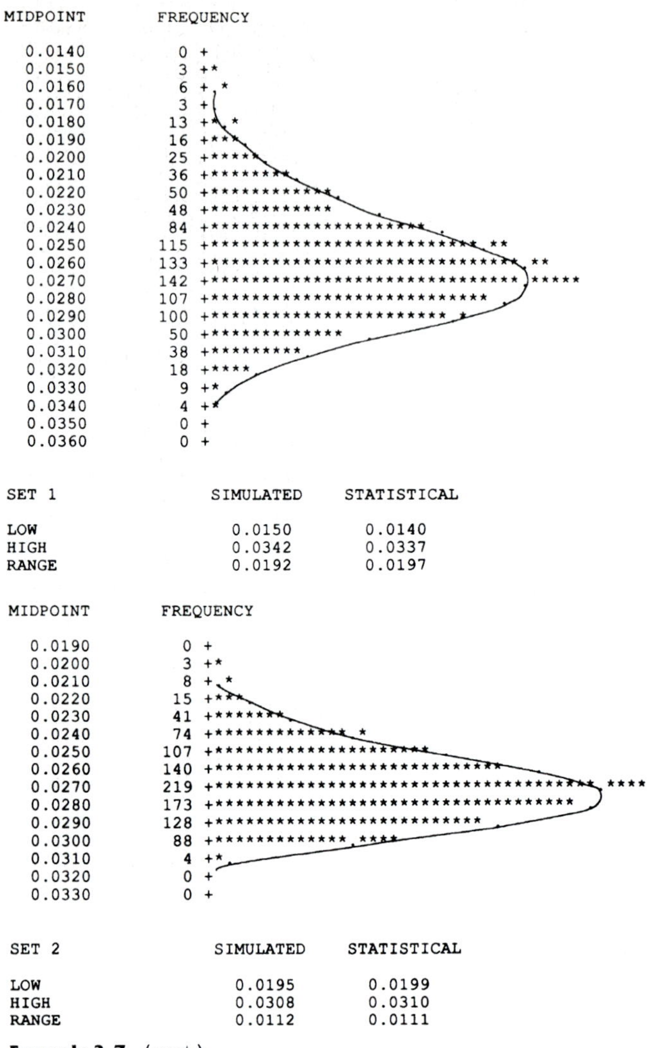

MIDPOINT	FREQUENCY
0.0140	0
0.0150	3
0.0160	6
0.0170	3
0.0180	13
0.0190	16
0.0200	25
0.0210	36
0.0220	50
0.0230	48
0.0240	84
0.0250	115
0.0260	133
0.0270	142
0.0280	107
0.0290	100
0.0300	50
0.0310	38
0.0320	18
0.0330	9
0.0340	4
0.0350	0
0.0360	0

SET 1	SIMULATED	STATISTICAL
LOW	0.0150	0.0140
HIGH	0.0342	0.0337
RANGE	0.0192	0.0197

MIDPOINT	FREQUENCY
0.0190	0
0.0200	3
0.0210	8
0.0220	15
0.0230	41
0.0240	74
0.0250	107
0.0260	140
0.0270	219
0.0280	173
0.0290	128
0.0300	88
0.0310	4
0.0320	0
0.0330	0

SET 2	SIMULATED	STATISTICAL
LOW	0.0195	0.0199
HIGH	0.0308	0.0310
RANGE	0.0112	0.0111

Example 3-7. (cont.)

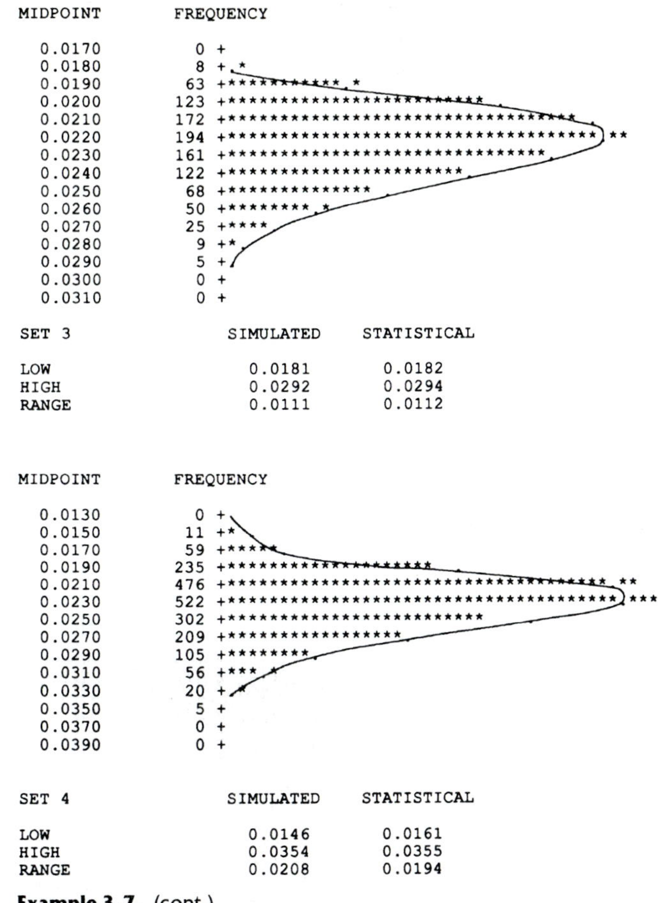

MIDPOINT	FREQUENCY
0.0170	0
0.0180	8
0.0190	63
0.0200	123
0.0210	172
0.0220	194
0.0230	161
0.0240	122
0.0250	68
0.0260	50
0.0270	25
0.0280	9
0.0290	5
0.0300	0
0.0310	0

SET 3	SIMULATED	STATISTICAL
LOW	0.0181	0.0182
HIGH	0.0292	0.0294
RANGE	0.0111	0.0112

MIDPOINT	FREQUENCY
0.0130	0
0.0150	11
0.0170	59
0.0190	235
0.0210	476
0.0230	522
0.0250	302
0.0270	209
0.0290	105
0.0310	56
0.0330	20
0.0350	5
0.0370	0
0.0390	0

SET 4	SIMULATED	STATISTICAL
LOW	0.0146	0.0161
HIGH	0.0354	0.0355
RANGE	0.0208	0.0194

Example 3-7. (cont.)

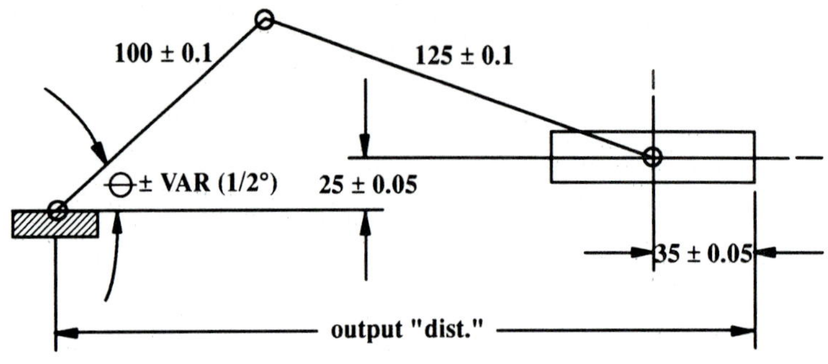

	θ°	LOW	HIGH	RANGE	MEAN	MAJ.CONTRIB %
		---STATISTICAL---				
DIST1	0	257.21	257.73	0.5246	257.47	θ 61
DIST2	15	256.30	256.89	0.5878	256.79	θ 74
DIST3	30	243.48	244.69	1.2133	244.08	θ 95
DIST4	45	221.20	222.93	1.7330	222.06	θ 98
DIST5	60	192.77	194.79	2.0194	193.78	θ 98
DIST6	75	163.35	164.38	2.0250	163.36	θ 98
DIST7	90	134.12	135.91	1.7849	135.01	θ 97

Example 3-8. Linkage simulation (VAR uniform, all others normal)

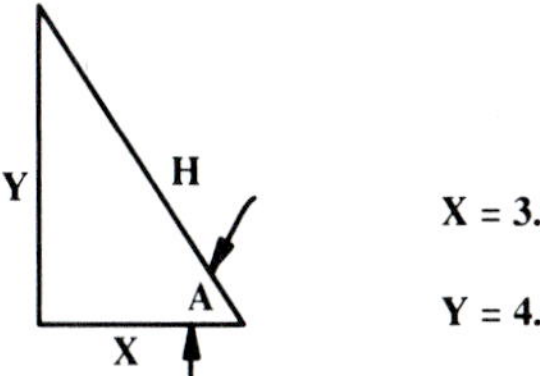

$X = 3.0 \pm 0.5$

$Y = 4.0 \pm 0.3$

Determine values for *A* and *H*

Equations:

$$A = \tan^{-1} \frac{Y}{X} \qquad H = \sqrt{X^2 + Y^2}$$

By the RMS method:

$$\frac{\delta A}{\delta X} = \frac{-Y}{X^2 + Y^2} \qquad \frac{\delta H}{\delta X} = \frac{X}{\sqrt{X^2 + Y^2}}$$

$$\frac{\delta A}{\delta Y} = \frac{X}{X^2 + Y^2} \qquad \frac{\delta H}{\delta Y} = \frac{Y}{\sqrt{X^2 + Y^2}}$$

From Equation (3-4):

$$T_A = \sqrt{\left(\frac{\delta A}{\delta X} T_X\right)^2 + \left(\frac{\delta A}{\delta Y} T_Y\right)^2} \qquad T_H = \sqrt{\left(\frac{\delta H}{\delta X} T_X\right)^2 + \left(\frac{\delta H}{\delta Y} T_Y\right)^2}$$

$$T_A = \frac{\sqrt{Y^2 T_X{}^2 + X^2 T_Y{}^2}}{X^2 + Y^2} \qquad T_H = \sqrt{\frac{X^2 T_X{}^2 + Y^2 T_Y{}^2}{X^2 + Y^2}}$$

Example 3-9.

$T_A = \sqrt{\frac{16\,(.25) + 9\,(.09)}{9 + 16}}$	$T_H = \sqrt{\frac{9\,(.25) + 16\,(.09)}{16 + 9}}$
$T_A = \sqrt{\frac{4.81}{25}}$	$T_H = \sqrt{\frac{2.25 + 1.44}{25}}$
$T_A = \pm.0877$ radians	$T_H = \sqrt{.1476}$
$T_A = \pm 5.02°$	$T_H = \pm.3842$

VARIABLE # 1	SIDE H		SIMULATED	STATISTICAL
SAMPLE SIZE	1000			
NOMINAL	5.0000	LOW	4.6624	4.6536
MEAN	5.0001	HIGH	5.4022	5.3466
STD DEV	0.1266	RANGE	0.7398	0.6930
SKEWNESS	0.0254	CAPABILITY INDEX		
KURTOSIS	-0.03124	SHIFT FROM NOM	0.0001	

VARIABLE # 2	ANGLE A		SIMULATED	STATISTICAL
SAMPLE SIZE	1000			
NOMINAL	53.1301	LOW	47.9772°	41.1897°
MEAN	53.0904	HIGH	57.7764°	57.9911°
STD DEV	1.6334	RANGE	9.7992°	9.8014°
SKEWNESS	0.0329	CAPABILITY INDEX		
KURTOSIS	-0.1725	SHIFT FROM NOM	-0.0397	

Example 3-9. (cont.)

Example 3-9 examines a simple triangle to determine two-dimensional relationships by the two methods.

SUMMARY

An understanding of the frequency distribution of a process or parameter opens the door to statistical analysis methods. Such methods can lead either to reduced stack-ups (assembly tolerance) or increased component tolerance. The Bender and Gilson factors allow a probability calculation to be more conservative when process controls are not adequate. Expected variation values can be separated from full probability stack-up values when a significant fixturing effect exists. Tight tolerance requirements can be controlled by selective assembly and shimmed assembly techniques. Monte Carlo simulation provides a method of analyzing complex assemblies and a means to deal with frequency distribu.oj offtions that are not normal.

Chapter 4

Statistics in Manufacturing

When we first introduced statistical methods, we noted that an assumption was made that the parts had variation which was described by the *normal distribution*. In fact, the underlying assumptions are more complex. They include these facts:

1. All distributions are normal. As seen in Figure 3-2, distributions are frequently rectangular, skewed, and/or other irregular shapes. Inspection processes may remove parts resulting in a truncated distribution. The parts may come from several sources and have distributions which overlap, resulting in multi-modal distributions. Fortunately, non-normality may not seriously effect a stack-up, especially if there are many components with nearly equal variation. Figure 4-1 illustrates a worst case scenario: how parts with rectangular variation distributions produce assemblies with distributions that approach normal.

2. The parts must come from a controlled process. If a part is not in statistical control, results can be unpredictable. It is exactly this sort of process that may also be influenced by excessive adjustments or the sorting of defective parts.

3. The process spread for the assembly exactly equals the specified tolerance. Figure 4-2a shows this condition where the width of the distribution curve matches the tolerance limits. This requirement is practically impossible to meet and can significantly effect the validity of a statistical calculation.

4. The process average matches the design mean. Figures 4-2a and 4-2b fulfill this requirement, however, Figure 4-2b violates item 3. This condition is frequently encountered because a common manufacturing goal is to work within 75% of the specified toler-

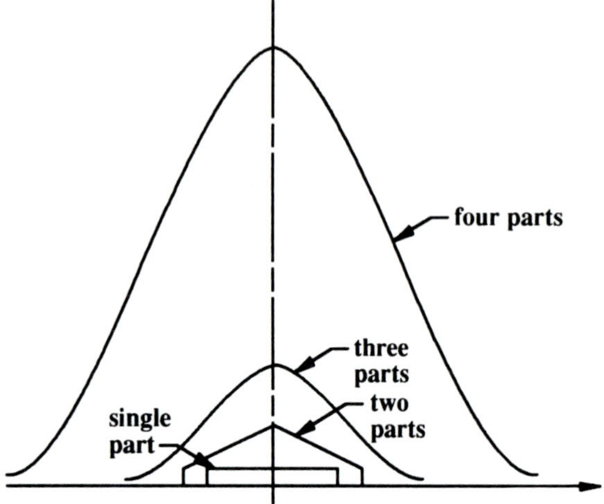

Figure 4-1. Summation of rectangular distribution

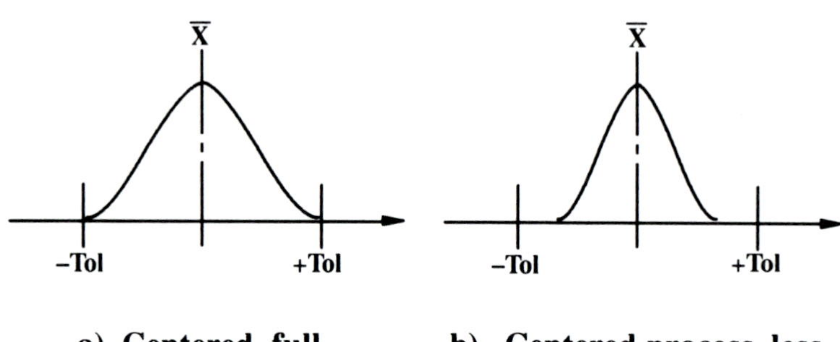

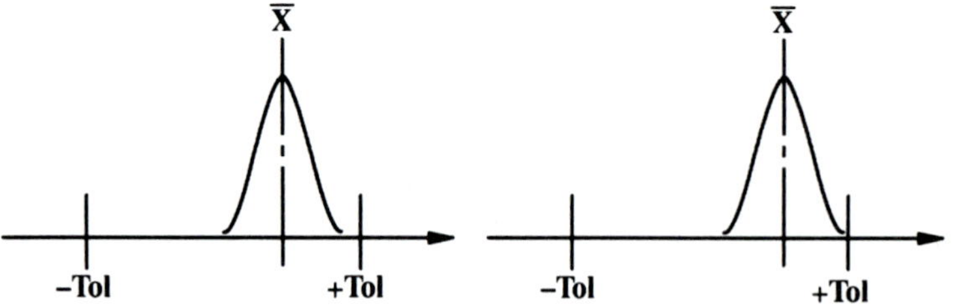

Figure 4-2. Process centering

ance. One of the serious problems with this requirement of centering the process is that many metal forming and machining processes, where the tool wears, have traditionally been tooled to produce parts at one end of the tolerance spectrum and allowed to wear toward the other. This results in distributions like Figure 4-2d at the beginning and end of the tool life. Processes with distributions of this nature are unsatisfactory because of the high concentration of parts at the extreme of the tolerance. Since most of the parts are in the *tail area* of the expected distribution of Figure 4-2a, the probability of selecting parts that will not assemble is dangerously high.

5. The component variations are independent. This requirement may not be a problem in many cases, however, certain industrial practices or processes can seriously violate this principle. The simple part in Example 3-1 is a classic example of parts that come from one piece of material and thus have common thickness or from sequential operations of the same tool, again, having little or no variation. Parts from a rework operation can seriously affect statistically based operations as they may have been out of tolerance and reworked to be just within the tolerance range. The majority of the parts may be clustered at a one-dimensional extreme.

The answer to these conditions and their possible deviations lies in process control. This works best in companies that practice simultaneous engineering discipline, bringing their product and manufacturing engineers together to make the necessary tradeoffs so as to achieve process control and permit statistical dimensioning to work.

The first required step is to identify dimensions used in statistical analysis. The note, *statistical dimension* adjacent to the dimension and tolerance involved, has been used by several companies. Other companies have used a symbol defined in their own company standards to denote such dimensions. It seems appropriate for companies that use *critical characteristic notation systems* to include statistical dimension labeling within that system. Such a symbol system is discussed in Chapter 5. Whatever the means, the first step lies in identifying where the process controls are required.

Once identified, responsibility for the statistical parameters shifts to the manufacturing and quality engineers. Referring to the previously listed process control requirements, some potential steps that can be taken include the following.

1. *Normality*—Review the process or similar process distributions. Limit sources to one supplier. Prohibit sorting and rework unless under the control of a quality engineer.
2. *Controlled Process*—Apply *statistical process control* (SPC) to the process, at least in the early stages of production.
3. *Natural Tolerance Span*—Check process variation statistically.
4. *Process Centering*—Prohibit tooling practices that result in working from one limit to another. Build adjustment into fixtures and tools so that they can be centered at their mean. Control purchase of materials so that the material is not deliberately bought thin or sold thick.
5. *Independent*—Make sure processes do not introduce dependance. Again, control rework.

Following such practices, one product, a relay, was assembled with 95% of more than 1 million units meeting specification without adjustment. In this unit, component tolerances averaged three times those allowed by linear tolerance.

Process Variation Analysis

Many process problems can be analyzed by plotting a histogram from available data and studying the distribution. Figure 4-3[1] shows many possible distributions. Their causes are noted below:

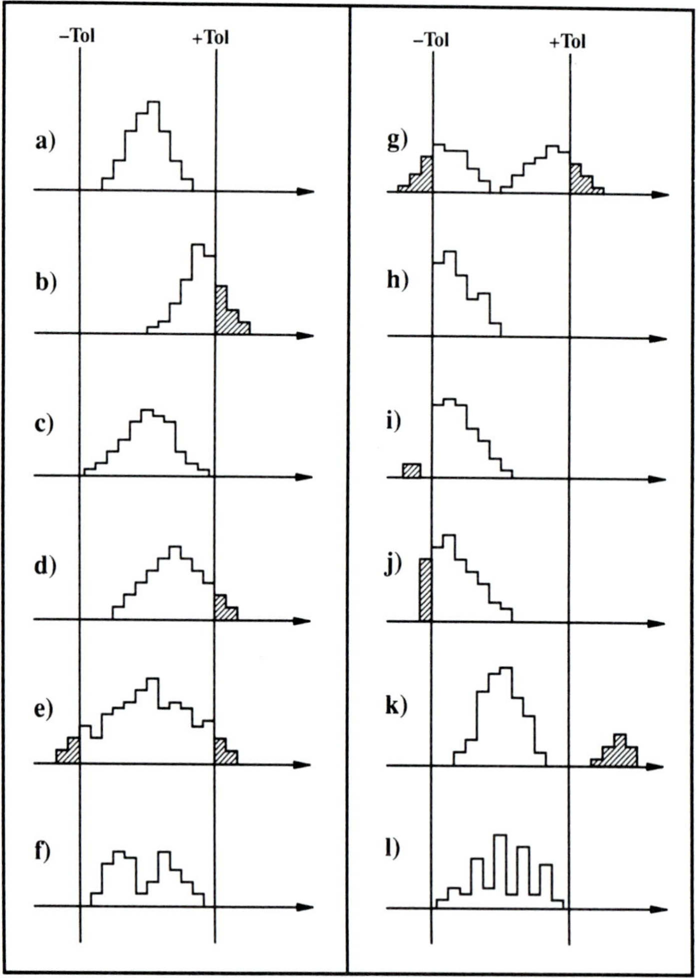

Figure 4-3. Use of histogram as analysis tool

[1]This illustration and those similar in format have been used in statistical training in the automotive industry for several years. The author has been unable to locate the original, but acknowledges the anonymous author.

a. Good process.

b. Mean value shifted toward high limit.

c. Acceptable, but the ends of distribution crowd the tolerance limits. A small percent of parts may be beyond the limits and any shift in mean value will produce out of specification parts.

d. Same as distribution 3 above with the mean shifted toward the high limit producing some out of specification parts.

e. The distribution spread is too wide, producing out of specification parts at both limits.

f. Parts are coming from two distributions—two machines or two sources, and so on.

g. Same as distribution 6 above with parts out of specification at both limits.

h. Skewed process: This occurs many places where negative values are illogical (e.g., a flatness measurement).

i. The result of sorting and not finding all of the bad parts. Classically, 100% inspection is only 80% effective.

j. The result of sorting and a relaxation of the lower limit in order to accept out-of-specification parts, sometimes called a *salvage limit.*

k. A classic problem: The small distribution represents the parts used for set-up. A high mean was corrected but the parts were not scrapped and became mixed with the good production parts.

l. Dead bands created by lack of discrimination in the measurement device or from two or more persons taking measurements. For example, one person may round off all values to whole numbers thereby creating gaps in the *fractional* areas.

Figure 4-4 illustrates another means of viewing data in order to gain insights into process trends. These insights are gained by plotting the observations in sequence so that time trends can be observed. Figure 4-4a shows data which has a mean that is trending higher with time. Many of the available computer programs can make such a determination. Figure 4-4b shows a data set where the process mean was obviously shifted at observations numbered 20 and 34. Such observations should be substantiated by setup or process logs where available.

STATISTICAL PROCESS CONTROL

Statistical process control (SPC) is built upon use of *process control charts* as a means of evaluating and controlling the process at hand. While their use as a control device is of the utmost importance, and worthy of extensive detail coverage, our focus here is upon their use as a variation management tool. In that context, their value lies in determining when a process is in statistical control so that process capability may be established. Knowledge of process capability is needed to close the loop of *specification* to *process* to *product* so that the analytical and specification methods used reflect real world conditions and, consequently, validate our variation management techniques.

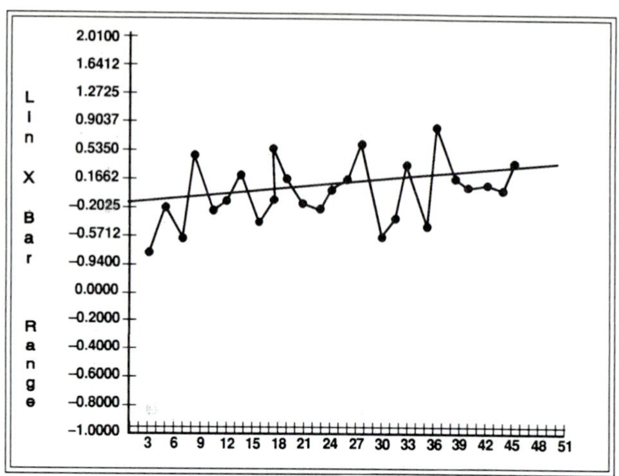

a) Data indicates a time trend

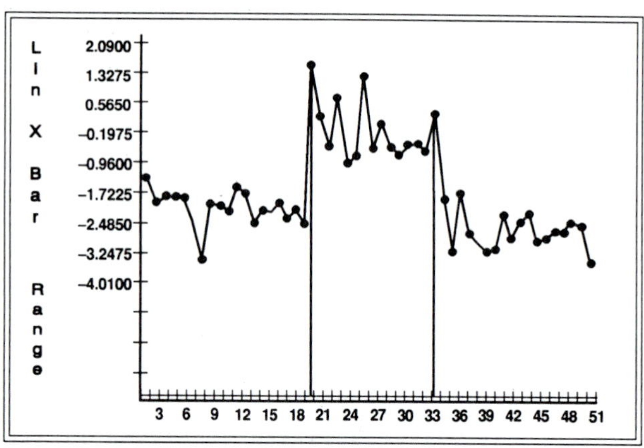

b) Data indicates change in mean

Figure 4-4. Sequenced data

$\overline{X}$ AND *S* CHARTS

Figure 4-5 illustrates a typical $\overline{X}$ and S chart. $\overline{X}$ values represent the average X value (hence $\overline{X}$) of a small subgroup (typically 3 to 5 parts), plotted periodically. Subgroups are used as their averages tend toward normal distribution, even though the parameters that they represent may come from frequency distributions which are not normal. The control chart has *upper* (UCL) and *lower* (LCL) *control limits* based upon a prior group of samples.

When used as a control device, a blank chart with the established limits is made for the process to be observed. Subgroups are selected and measured periodically (say each half hour). They are then plotted yielding a chart like the upper chart of Figure 4-5. If the

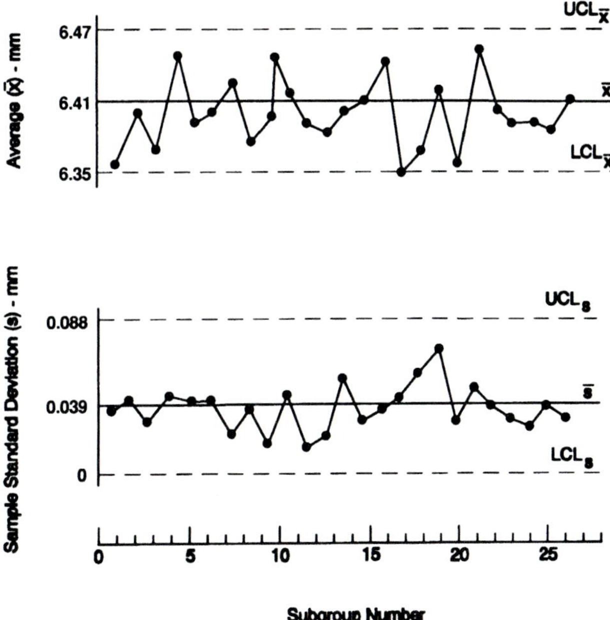

Figure 4-5. $\overline{X}$ and S chart

plotted points lie within the control limits and do not exhibit trends or other conditions[2] which imply nonconformance to expected performance, the process is said to be in statistical control.

When the process is not in control, the special causes (sources of the problem) must be found and appropriate action taken to eliminate them. Note that SPC does not define causes, but rather, only signifies that special causes exist. When all of the special causes have been eliminated, only the common causes remain. Many excellent quality control books are available for further information on the use of control charts.[3]

The lower chart in Figure 4-5 represents the variation of the standard deviation (S) of the subgroups plotted. Not only must the average value ($\overline{X}$) of a subgroup conform to expectations, but also the spread (S) of the individual parts must also perform predictably.

The $\overline{X}$ and S chart has been used less frequently than the $\overline{X}$ and R chart. R denotes the range of observed values for each subgroup. Use of this type of chart dates from a period lacking modern analytical and process computers with the ability to calculate the S value with ease. Computer use has allowed S values to be calculated more easily.[4] Recognizing the many computer aids available, this author recommends that $\overline{X}$ and S Charts be used in lieu of the $\overline{X}$ and $\overline{R}$ charts generally used.

[2] See the Appendix for various available programs.

[3] One such reference is *Quality Control*, by Dale H. Besterfield, Prentice Hall, Englewood Cliffs, NJ, 1986.

[4] Hancock, W. M., and Yang, K. *Advances in Statistical Process Control*, The University of Michigan Quality Assurance Conference, Traverse City, MI, August 1988.

Tampering

It is important to note that when a process is in statistical control, undisciplined adjustments (perhaps directed by management) will increase the process variability as shown in Figure 4-6. Often, the first step in resolving a process problem is to stop all adjustments and to determine the results of the stable process. If it is in statistical control, the variation is from common causes and can be reduced only by a system change such as a product or process redesign.

Relationship to Tolerance Specification

Process control limits are for averages and result from the process observed, while tolerance specifications are for individual values and are established to ensure product function. Figure 4-7 contrasts the distribution of $\overline{X}$, the distribution of X (which has a larger span) and the tolerance specification. The process and, hence its control limits, is independent of the tolerance specification, however, certain limits exist:

where: T = Tolerance

n = Subgroup size

$\sigma_x = \dfrac{\sigma}{\sqrt{n}}$ for normal distributions

$\dfrac{T}{3\sigma} > 1.33$ from the following section

then as an upper limit:

$$\sigma_x < \frac{T}{4\sqrt{n}}. \tag{4-1}$$

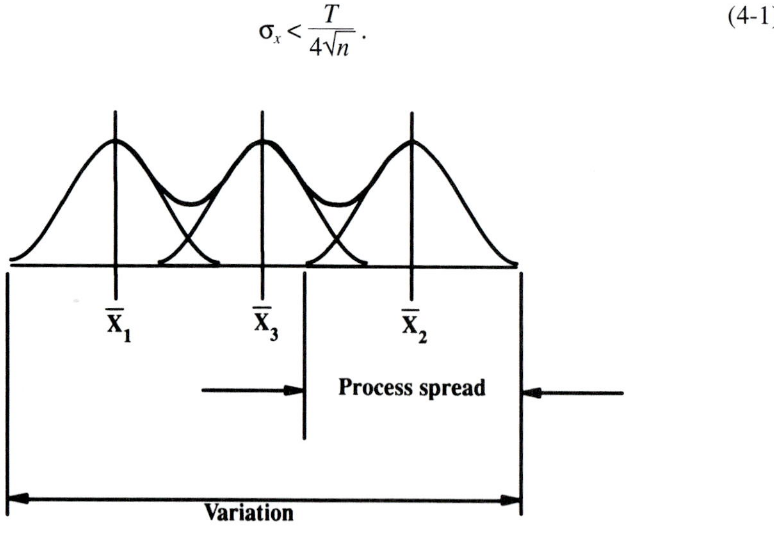

Figure 4-6. Tampering

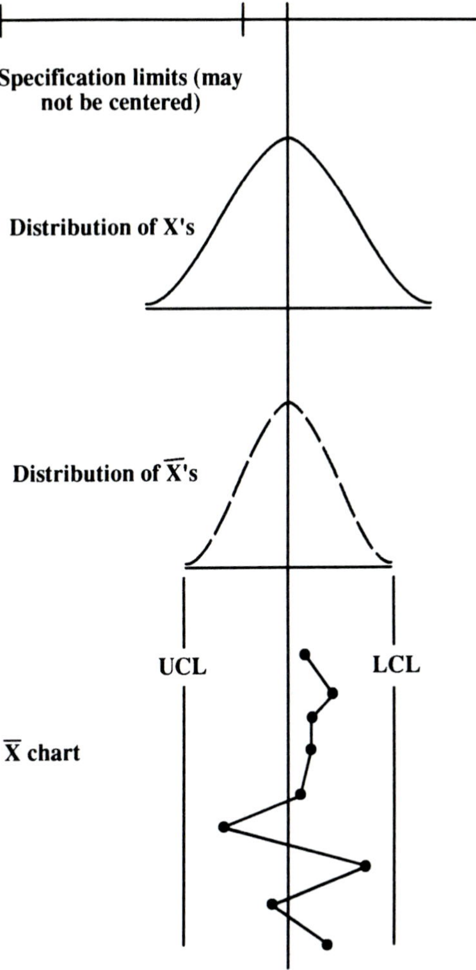

Figure 4-7. Relationship of $\overline{X}$ to tolerance specification

PROCESS CAPABILITY

When a process is first observed, typically, it will have special causes present. The subgroups will vary widely and the process will lack control as shown in Figure 4-8a.[5] As special causes are found and eliminated, the subgroups form a consistent, predictable distribution as shown in Figure 4-8b. The process is then said to be in statistical control.

Now the process output can be compared to the specifications in order to determine

[5]This illustration is from "Continuing Process Control and process capability Improvement," a copyright publication of the Quality Education and Training Center of the Ford Motor Co. It is reproduced with their permission.

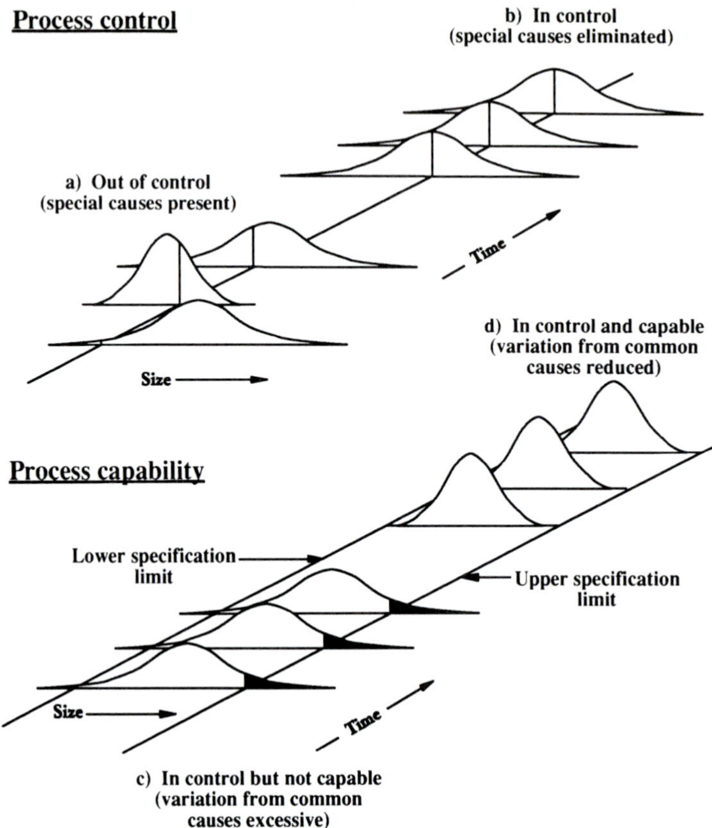

Figure 4-8. Process control and capability

the capability of the process. Figure 4-8c shows a process where the upper and lower tails of the distribution of individual parts lie outside of the specification limits, hence, the process is not capable. Refinement of the process ultimately results in reduced variation so that nearly all of the process output now lies within the specification limits as shown by Figure 4-8d. Such a process now demonstrates the capability to meet the specification limits.

MEASURES OF PROCESS CAPABILITY

Two *measures of process capability* are frequently used. The first is the C_p ratio. C_p is the ratio of product tolerance divided by the process spread, (see Figure 4-9a). To be capable, a process must have a C_p value ≥ 1. A value of 1.33 is frequently used as a goal. This value occurs when the process spread is 75% of the specification limits. The logic of having a goal value larger than 1.0 lies in the fact that many processes have stable repetitive variation but suffer from drift of the mean ($\overline{X}$) value. This can result in producing out of specification parts as shown in Figure 4-9b although the process has capability (acceptable varia-

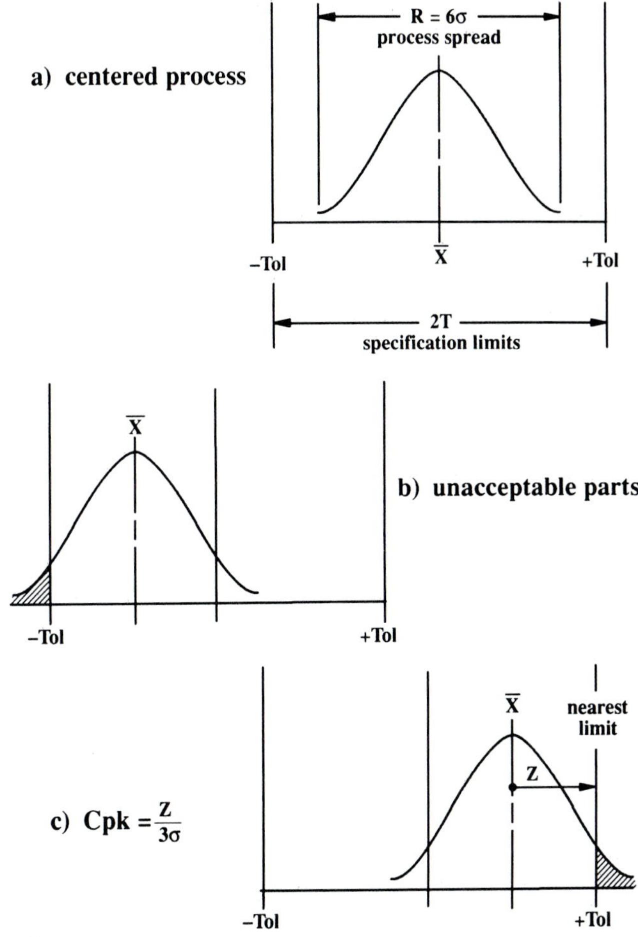

Figure 4-9. Process capability and centering ratios

tion). A similar result can occur when the mean remains stationary and the spread increases as can occur with tool wear.

Figure 4-9c shows how this centering effect is measured by the C_{pk} parameter. Here the *available tolerance* , the distance from $\overline{X}$ to the nearest specification limit (denoted Z), is divided by one-half the process spread and designated C_{pk}. Like the C_p parameter C_{pk} values ≥ 1 are required for process capability. The 1.33 goal is also frequently used for C_{pk} values.

In practice, both C_p and C_{pk} values are used together to describe the process potential (C_p) and its actual performance including the effect of centering (C_{pk}). Figure 4-10 shows how C_p and C_{pk} values are necessary to fully define a process. In this case, both of the examples have a C_{pk} value of 1.0 while the C_p value differs. Although the choice of processes is not obvious, the process with the highest C_p value would normally be selected, especially if it has a possibility of being recentered.

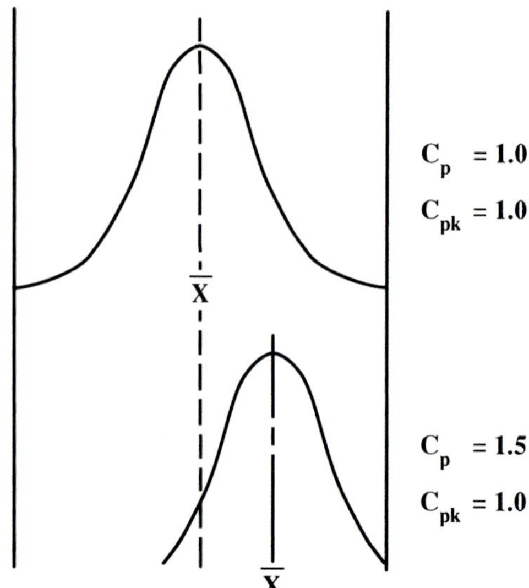

Figure 4-10. Use of C_p and C_{pk} values

Figure 4-11 indicates how the C_{pk} value changes as the process trends away from the design mean. Thus, we see that C_{pk} and C_p match when the process is centered as in Figure 4-11a. *Process capability* is maintained until the C_{pk} value equals 1.0 as in Figure 4-11b. As C_{pk} becomes less than 1.0, unacceptable parts begin to be produced as shown in Figure 4-11c. A C_{pk} index may also have a zero or negative value. Superficially, this sounds unreasonable, but further examination shows that such values give us insight about the centering accuracy of the process. Zero C_{pk} values (or values very near zero) indicate that the process mean coincides with one of the specification limits, as shown in Figure 4-11d, resulting in 50% of the population being out of specification. Negative C_{pk} values indicate the process mean lies outside the specification limit, as shown in Figure 4-11e, while in Figure 4-11f, 100% of the population is out of specification when C_{pk} reaches –1.0. The higher the negative value, the further the process mean lies beyond the specification limits.

Six Sigma Capability

A few advanced corporations are now developing their capability based upon a plus or minus 6 sigma tolerance range (12 sigma range). These systems require:

$C_p \geq 2.0$
$C_{pk} \geq 1.5$
$C_c \leq 0.25$ (This parameter is described in the section on Process Centering)

A study of these parameters will show that the $\pm 6\ \sigma$ tolerance yields a C_p of 2.0. The targeting requirement equates to $\pm 1.5\ \sigma$, retaining 4.5 σ within the tolerance limits yielding a C_{pk} of 1.5. A 4.5 σ coordinate on a normal distribution equates to a defect level of about 3.4 parts per million.

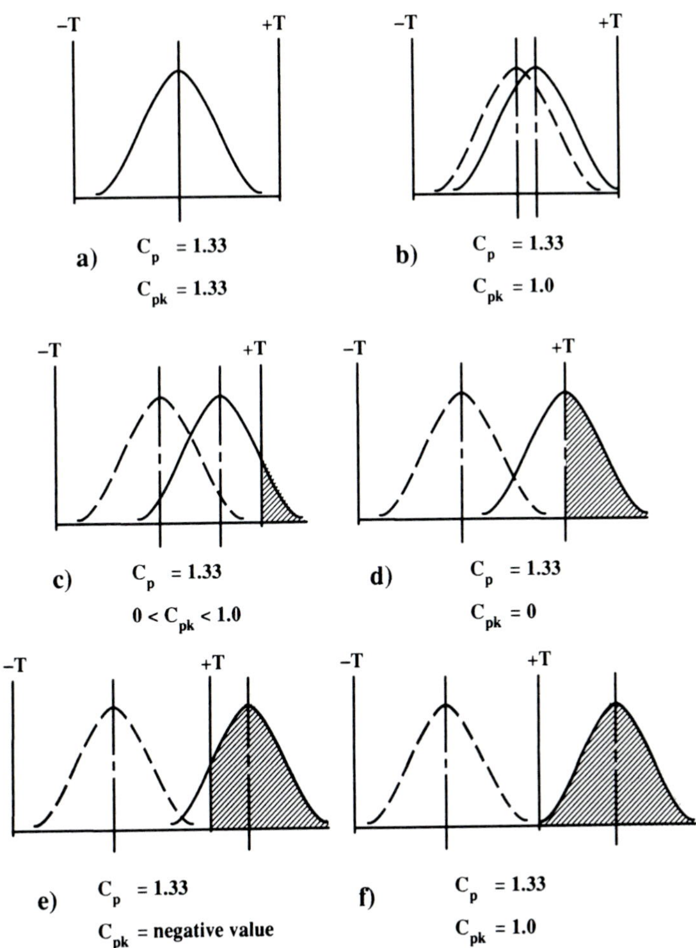

Figure 4-11. C_{pk} trends

Process Centering

To support characteristics which require centering, a new parameter C_c (similar to C_p and C_{pk})[6] is used to limit the centering error as shown in Figure 4-12. The major benefit of capping the C_c value is that the expansion factors used with statistical calculations to compensate for uncontrolled centering errors can be handled in a more definitive manner. Although it is possible to assume equal component tolerances and calculate an expansion factor, it is better to formulate the assembly relationship based upon the potential mean shifts and alter the stack-up calculation method. Two cases which share a common formula are shown here.

[6]Other designations have been used to identify this parameter elsewhere in the literature. One such designation is *K*.

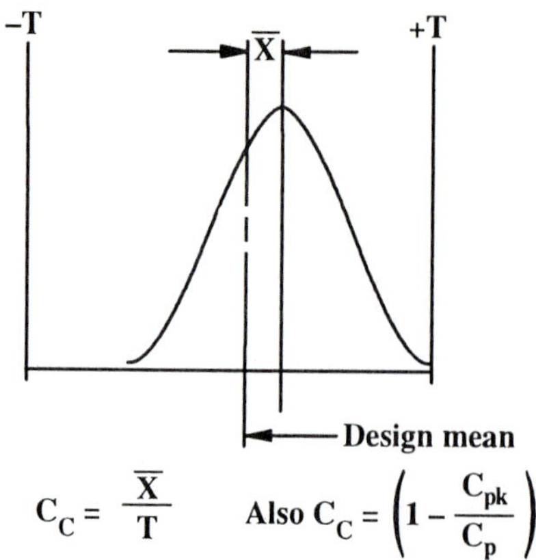

$$C_C = \frac{\overline{X}}{T} \qquad \text{Also } C_C = \left(1 - \frac{C_{pk}}{C_p}\right)$$

Figure 4-12. Measurement of process centering

For 1 part per 1,000 defective in assembly (± 4 sigma):

$$C_p \geq 1.33,\ C_{pk} \geq 1.0, \text{ and } C_c \leq 0.25$$

or, for 3 parts per 1,000,000 defective in assembly (± 6 sigma)

$$C_p \geq 2.0,\ C_{pk} \geq 1.5, \text{ and } C_c \leq 0.25:$$

$$T_{ASSAY} = C_c(\Sigma t_i) + (1 - C_c)\sqrt{\Sigma t_i^2}$$

The author strongly recommends this analysis approach and the 6 sigma philosophy. If 6 sigma capability has not yet been reached, the fact that a common calculation can be used for both the 4 and 6 sigma systems allows you to progress from the lower to the higher quality without altering the engineering calculations or specifications. Example 4-1, using the data of Example 3-2, illustrates a typical analysis.

Non-normal Data

It is obvious that the definitions of process capability are based upon *normal* distributions. Equally obvious is the fact that measured data is not always normal. In practice, this is not much of a problem as the capability calculations look only to end points (± 3 σ) of a distribution. Many statistical computer programs fit a *Pearson curve* to *non-normal data*. It is possible to determine the end points for these curves that directly correspond to the normal (± 3 σ) end points. Use of these end points in a capability index calculation frees it from the assumption of normality.

A final note on capability indexes: Although they appear precise (2 decimal places for the most part), they are influenced by sampling variation, by the fact that no process is ever fully in statistical control, and by issues relating to normality. Hence, capability indexes should always be considered to be approximations.

RSS CALCULATIONS						
SUBJECT: DISC BRAKE		ORIGINAL DATE: ANALYST: JVL		REVISED	PART NO.: PAGE:	
NO	DESCRIPTION	CONTRIB. FACTORS RANK	%	± TOLERANCE LINEAR	SQUARED	MEAN (MEAN SHIFT)
1.	PISTON			0.1	0.01	-48.0
2.	CALIPER WALL			0.1	0.01	-13.5
3.	CALIPER DEPTH			0.1	0.01	+125.0
4.	OUTER LINING			0.5	0.25	-14.2
5.	ROTOR			0.4	0.16	-30.0
6.	INNER LINING			0.5	0.25	-17.8
	SUM			1.7	0.69	1.5
	.25 ΣT			0.425		
	RSS	√			0.83	
	.75 RSS			0.623		
	VARIATION = .25 ΣT + .75 RSS			1.048		

Example 4-1. Calculated Mean Shift

SUMMARY

The use of statistical methods in manufacturing provides new prospectives to the control and analysis of processes that could not be attained by considering individual parts.

When statistical dimensions are used, they must be so labeled. Then appropriate processes need to be reviewed for:

1. Normality
2. Statistical control
3. Process spread versus tolerance specification

4. Centering
5. Independence

Some of the appropriate responses for variation control includes:

1. Limited or single sources
2. Control of rework and sorting
3. Adjustable tooling
4. Independence built into processes

Analysis of histograms can identify the causes of variability.

Statistical process control is used both to establish process control and as a subsequent monitoring device. $\overline{X}$ and S charts are recommended versus $\overline{X}$ and R charts.

Process capability measurements are stated in terms of C_p and C_{pk} ratios. C_p values greater than 1.0 (2.0 preferred) indicate process capability (acceptable variability). C_{pk} values greater than 1.0 (1.5 preferred) indicate acceptable centering as well. Both values are needed to describe a process.

Limiting process centering errors by the C_c parameter makes it possible to calculate the mean shifts that affect an assembly and to avoid the use of expansion factors in stack-up analyses. Additional process controls are needed for this parameter as improved C_p values may allow C_c to increase.

Chapter 5

Introduction to Geometric Tolerancing

Geometric dimensioning and tolerancing, or GD&T, is a method of conveying component tolerances through a symbolized notation system. Its merit, when compared to older dimension systems, lies in its ability to clearly convey intended part functions. Figure 5-1 illustrates the ambiguity associated with size tolerances. There is no clear tolerance zone such as: ± 2.0 on the right side, ± 2.0 on the left side, or ± 1.0 on each side. It becomes increasingly difficult to understand how the tolerance is distributed as the number of dimensions or part complexities increase.

In Figure 5-2, the GD&T used clearly places a 2.0 tolerance band on the right side and locates it from the datums A, B, and C. It is this ability to clearly communicate such relationships that makes GD&T such a powerful tool. The unique feature of the system that allows it to clearly communicate relationships is through the use of *datums* (datums are discussed in depth in Chapter 6).

Because of the importance of datums, this author now routinely decodes the initials GD&T as meaning *geometric datums and tolerances.* Chapter 7 treats the various classes of tolerance in detail. This chapter will focus on the symbolized notation system and the concept of *critical size,* both of which affect datum and tolerance specifications. *Critical characteristics,* although not normally considered as part of a GD&T system, are discussed as they support the implementation of tolerance control via symbolized notation.

SYMBOLIZED TOLERANCE NOTES

If you were to compare an ordinary mechanical drawing from practically any industry with an artists illustration of the same component, it would become obvious that a very fundamental difference in technique exists. Careful observation indicates that the mechanical

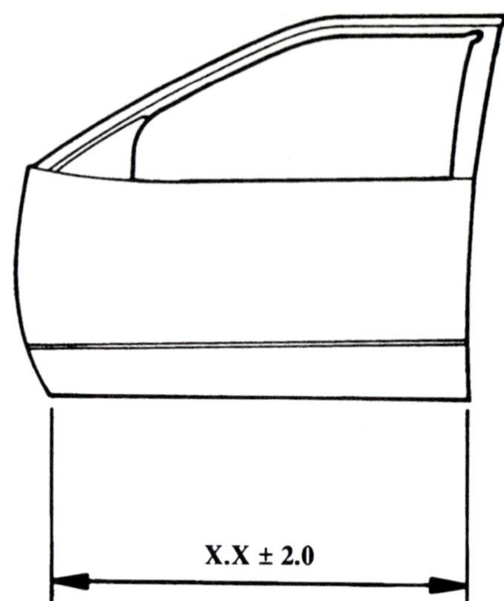

Figure 5-1. Old Way

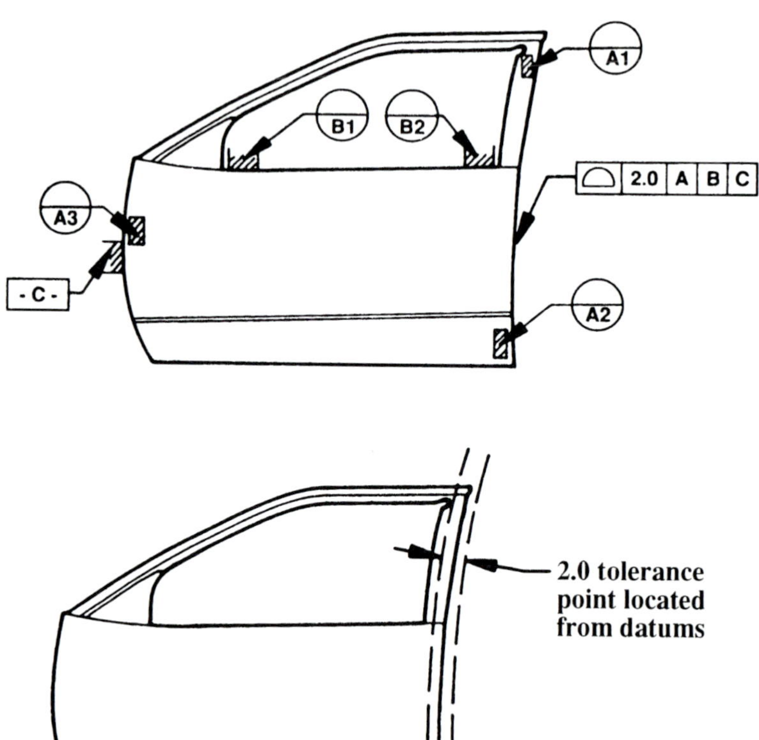

Figure 5-2. New Way

drawing entails the use of a carefully developed system of conventions chosen to convey precise meanings and to minimize drafting time. The engineer or draftsman may overlook this fact because of his intimate familiarity with the medium. Within different fields various degrees of symbolism exist, ranging from the placement of views to the use of symbolized notes (such as those used to define surface texture, welding requirements, and, more recently, geometric tolerances).

The complex form of many tolerance specifications makes it desirable to use symbolized notes, both to improve clarity and to reduce the time required to letter the notes on a drawing. Many corporations and government agencies prepare drawings that will be used in more than one country. When symbolized notes are used, the data may be more easily understood, being independent of the language in which the drawing notes are written. Coupled with these facts, the frequency with which complex tolerances appear on component drawings makes the symbolized tolerance specification very desirable.

In the United States, the symbol system used originated with the publication of MIL-STD-8 and is currently defined by ANSI Y14.5M-1982. Symbol usage has increased to the point where it is now nearly universal, the only exceptions being certain areas of deficiency in the standards which are discussed in the following sections. The geometric characteristic symbols defined by Y14.5 are shown in Figure 5-3.[1] These symbols are universally ac-

	TYPE OF TOLERANCE	CHARACTERISTIC	SYMBOL
FOR INDIVIDUAL FEATURES	FORM	STRAIGHTNESS	—
		FLATNESS	▱
		CIRCULARITY (ROUNDNESS)	○
		CYLINDRICITY	⌭
FOR INDIVIDUAL OR RELATED FEATURES	PROFILE	PROFILE OF A LINE	⌒
		PROFILE OF A SURFACE	⌓
FOR RELATED FEATURES	ORIENTATION	ANGULARITY	∠
		PERPENDICULARITY	⊥
		PARALLELISM	//
	LOCATION	POSITION	⌖
		CONCENTRICITY	◎
	RUN-OUT	CIRCULAR RUNOUT	↗
		TOTAL RUNOUT	⌰

Figure 5-3. Standard symbols for designating tolerance requirements

[1]This illustration is extracted from ANSI Y14.5M–1982 by the American Society of Mechanical Engineers, United Engineering Center, New York, NY and is reproduced with their permission.

cepted, both in English-speaking countries and in Europe (ISO standards), and define the type of tolerance specified.

In addition to these geometric symbols, there are symbols which are used to designate *critical size*. These symbols, often called *modifiers* because they modify the interpretation of the tolerance value, are shown in Figure 5-4a. The MMC and RFS symbol have been used for many years. The LMC symbol was first used by this author in *Fundamentals of Position Tolerance* in 1970 and was added to Y14.5 in 1982. The *virtual size* symbol was extracted from an obsolete General Motors tolerance standard.

Figure 5-4b shows *projected and fastener height modifiers* used to define tolerance zone extent and a datum modifier for non-rigid parts which require restraint. The projected height modifier is included in Y14.5 while the latter two designations are recommendations of this author. Figure 5-5 shows other accepted tolerance and geometry symbols.

The box used to identify basic dimensions is an indicator that general tolerances do not apply to the dimensions involved and that the GD&T is the controlling tolerance. Where CAD systems are used and the part geometry resides in a math data base, GD&T has priority over a general tolerance.

Feature Control Frame

All geometric and tolerance specifications include appropriate symbols and values to define the tolerance type, the tolerance value, datum references, and appropriate critical size modifiers within a rectangular frame called the *feature control frame* (shown in the various

SYMBOL	CRITICAL SIZE
Ⓜ	MAXIMUM MATERIAL CONDITION (MMC)
Ⓢ	REGARDLESS OF FEATURE SIZE (RFS)
Ⓛ	LEAST MATERIAL CONDITION (LMC)
Ⓥ	VIRTUAL CONDITION (VC)

a)

SYMBOL	MEANING
Ⓟ	PROJECTED HEIGHT
Ⓕ	FASTENER HEIGHT
Ⓡ	RESTRAINED DATUM

b)

Figure 5-4. Symbol modifiers designating critical size (a), and tolerance extent and datum restraint (b)

SYMBOL	MEANING	ABBREVIATION
⌀ (with leader arrow)	ALL AROUND	——
∅	DIAMETER	DIAM
(28.0)	REFERENCE	REF
[50.0]	BASIC DIMENSION	BSC
⌴	SPOTFACE COUNTERBORE	SFACE CBORE
⌵	COUNTERSINK	CSINK
↧	DEPTH	——
□	SQUARE	SQ
SR	SPERICAL RADIUS	SPHER R
S∅	SPHERICAL DIAMETER	SPHER DIA

Figure 5-5. Geometry and data symbols

examples of Figure 5-6). Current standards state that critical sizes must be noted for position tolerance specifications and that all other specifications imply an RFS critical size unless otherwise specified. Since the practice as to what is implied has changed over time, it is the recommendation of this author that critical size symbols always be used for features or datums of size. These relationships are far too important to risk misunderstanding and require very little time to document. When specified, the risk of change in implied interpretation is eliminated.

Figure 5-6 shows examples of tolerance specifications wherein the number of datums specified varies. As will become more evident in the chapter on datums, certain applications require a more complex set of datums than other applications. It is recommended that excess datums not be specified as this can lead to confusion with certain tolerance applications, particularly, *profile tolerance*. Also note that, as shown in Figure 5-6g, the placement of datums in the feature control frame has changed over time. Consequently, datums on older drawings may be shown preceding the tolerance value. The sequence of datums in the control frame designates the sequence (and implied importance) of datum use in manufacturing and inspection operations. Figure 5-6h illustrates a method used by Volvo of Sweden to identify the measuring method used for specific tolerances. This is done because measurement methods often involve some compromise from the strict interpretation of the tolerance specification.

Position Tolerance Values

Position tolerance values are now universally expressed as a diameter with the symbol "∅" proceeding the value, as shown in Figure 5-6c, or as a total value for other shapes. Previously, some standards such as the SAE and many companies expressed position tolerances by their radius value as shown in Figure 5-4g. Such specifications may be encoun-

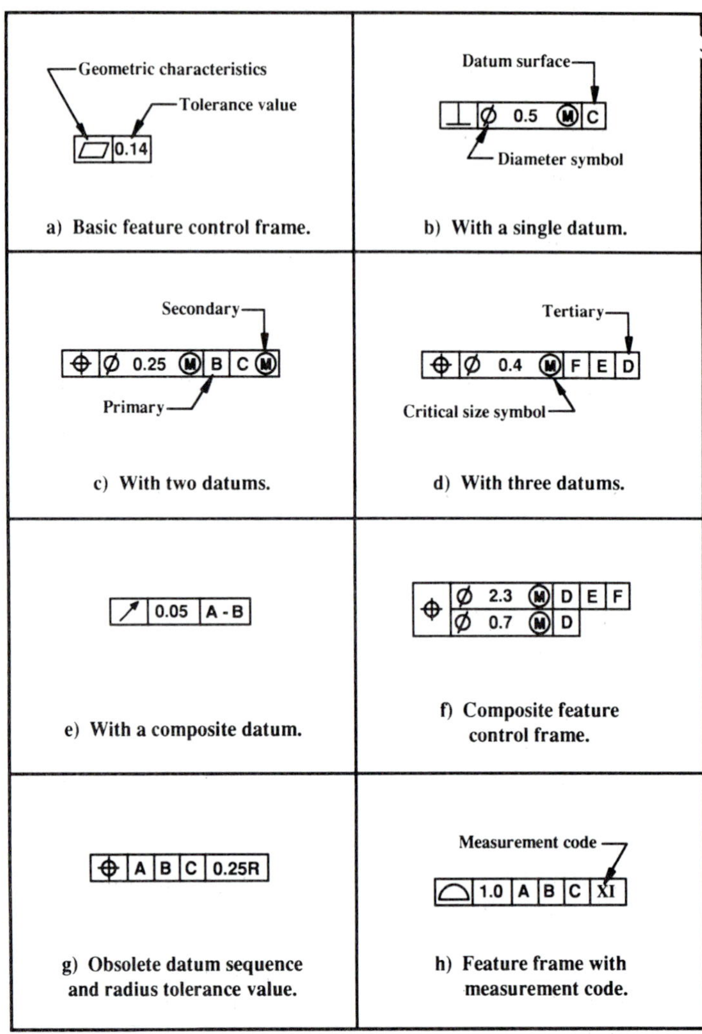

Figure 5-6.

tered on older drawings from time to time. Obviously, they can be multiplied by 2 to determine an equivalent diameter value. If doubt exists, interpret the value as a diameter or total as this is the smallest of the two possible interpretations. Also note that position tolerance specifications of this format were written during a period when MMC was implied as part of the specification.

Projected Height Designation

The basic symbol system outlined here does not describe *tolerance zone depth*. In the symbolized system, tolerance zone depth is assumed to be the same depth as the feature to which the tolerance control is applied. This leaves only the necessity of describing the pro-

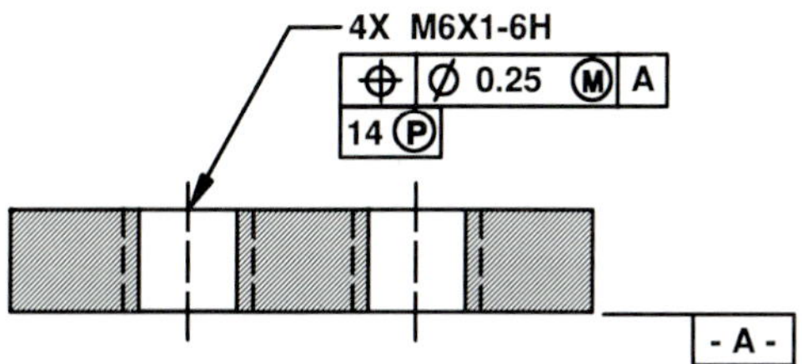

a) Projected height designation

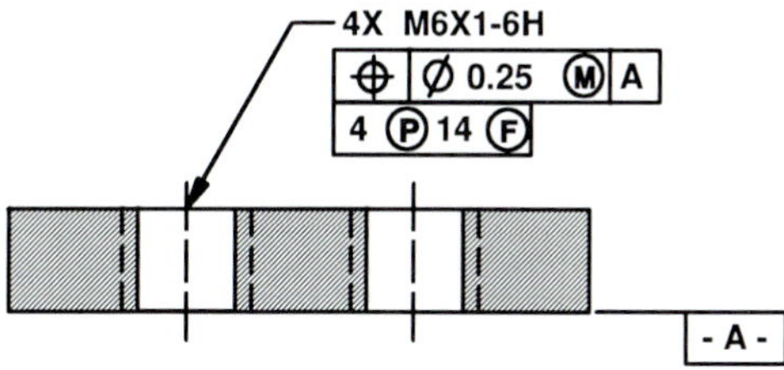

b) Projected height and fastener height designation

Figure 5-7. Tolerance extent symbols

jection or height of a tolerance zone that extends above the surface of a component as occurs with certain position tolerances. The *projected height symbol* of Figure 5-4b is included in Y14.5 while the *fastener height symbol* remains a recommendation of this author. Examples are shown in Figure 5-7.

DATUM DESIGNATIONS

In addition to being part of the tolerance specification, as shown in Figure 5-6, datum designations must be placed at or upon the features or surfaces they designate so as to indicate clearly the surfaces or features referenced in the tolerance specification. Y14.5 defines the two symbols in general use. The first, which is the basic designation symbol, is shown in Figure 5-8a. This symbol is used to identify planes, diameters, and the like. It is also used when a toleranced feature is designated as a datum, as shown in Figure 5-8d.

Figure 5-8b shows the accepted datum target symbol. The datum target may be a point, a line, or an area, and must be qualified by adequate description on the component drawing as shown in Figure 5-9. Square or round target sizes are included in the symbol. Figure 5-8c shows the previously used datum target symbol which may be seen on older drawings.

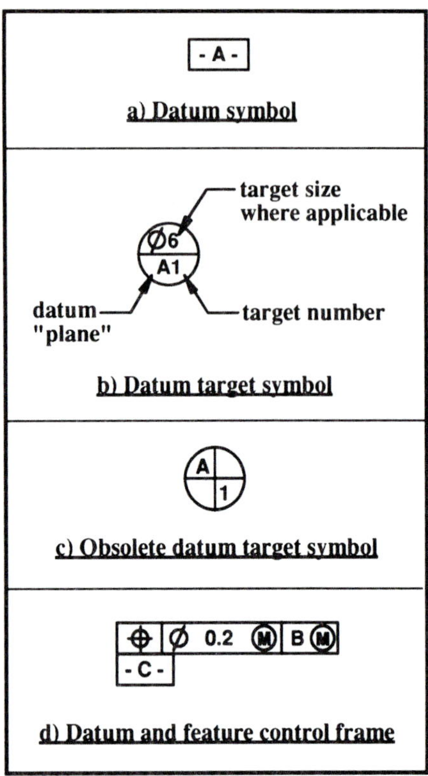

Figure 5-8. Datum symbols

CRITICAL SIZE

When features of size (diameters, holes, slots, etc.) are toleranced for attitude or position or used as a datum, the relation of the feature size to the position tolerance or to the datum function is significant. Therefore, methods have been developed to indicate the feature size at which the datum requirement or tolerance applies. This author calls these *critical sizes* because the datum relationship or tolerance applies at some specific size (i.e., the critical size). The various critical size categories, as shown in Figure 5-4a, are defined below:

Maximum Material Condition (MMC). The critical feature size for parts which require clearance to assemble. An MMC feature is one that is derived from its toleranced dimensions so that it contains the maximum amount of material with respect to any portion of the feature which will affect a position or form tolerance. For a hole, MMC represents the low-dimensional limit. For an outside diameter, MMC represents the high-dimensional limit.

Least Material Condition (LMC). The critical feature size for parts which require positioning so that they will assemble properly. An LMC feature is one that is derived from its toleranced features so that it contains the least amount of material with respect to any por-

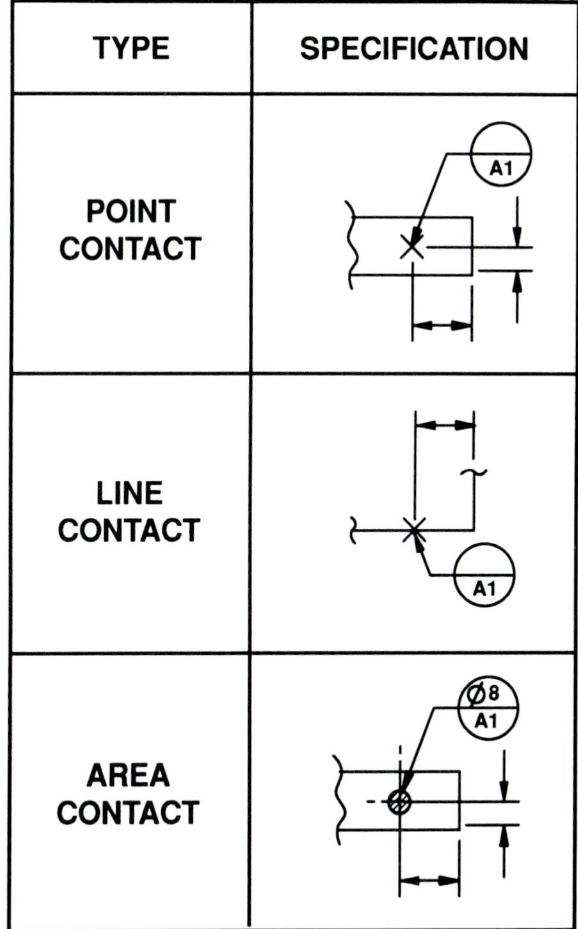

Figure 5-9. Datum symbols

tion of the feature which will affect a position or form tolerance. For a hole, LMC represents the high-dimensional limit. For an outside diameter, LMC represents the low-dimensional limit.

Regardless of Feature Size (RFS). Occupies a middle ground between MMC and LMC. It is associated with functions where alignment occurs as in the case of press fit components. RFS indicates that a feature is always at its critical size, hence, it can be conveniently thought of as indicating the center line of a feature.

Virtual Condition (VC). A composite of another critical size, typically MMC, and another tolerance such as squareness or position.

Current tolerance standards require that critical sizes be designated for features of size that are being toleranced or used as datums when position tolerances are applied. For other tolerances, the RFS condition is implied and the standards require only that critical sizes other than RFS be designated, when appropriate. This author recommends that critical sizes be designated for all features of size as those with little GD&T expertise will not know the implied requirements and the time required to complete the designation is minimumal. This will also provide protection against changes as to what is implied in various standards (which has occurred) and will prevent the tolerance notations from appearing in two forms where features of size are used with more than one type of tolerance on the same drawing.

CRITICAL CHARACTERISTICS

Of the many characteristics of a product defined by the dimensions and specifications on a drawing, only a few are critical to fulfilling the product's intended function. This is a *Pareto-type concept* that says that only a small number of contributors will dramatically effect the performance of a product. (The Pareto concept indicates that about 80% of the variation of an output is influenced by about 20% of the input characteristics.) Hence, *critical characteristic* is defined as a product characteristic for which reasonably anticipated variation (perhaps beyond the specified tolerances) could significantly affect product safety, function, or performance. This principle has long been recognized and many corporations now include symbols to designate critical characteristics (specifications, tolerances, and so on) on their product drawings. The basic idea behind their use is to communicate the product or manufacturing engineer's detailed knowledge to a downstream user, rather like highlighting certain sections of a written communication.

Critical characteristics have been used in the automotive industry for about 20 years. Their earliest use was to designate characteristics that affected performance of systems subject to Federal Motor Vehicle Safety Standards (FMVSS). Symbols used to designate safety and function characteristics are shown in Table 5-1. The *safety or compliance critical characteristics* identify parameters which require strict (100%) compliance to specifications. Their merit lies in their ability to identify critical areas that might otherwise be obscure as in the following cases:

1. Integrity of a weld in a steering wheel core where an improper or undersize weld could result in the loss of the wheel and the subsequent loss of vehicle control.
2. Minimum material thickness of a structural component (such as a Jack) that could fail under extreme load, if undersized.

The next category of critical characteristics to come into use was *functional characteristics.* These characteristics are similar to safety characteristics except that they relate to product failures with unsatisfactory, although less disastrous, consequences than safety characteristics. They usually require statistical compliance.

Critical characteristic symbols have little merit without the enforced reactions found in corporations that practice their use. The appropriate reactions by manufacturing and quality staffs for safety and function characteristics include such items as:

1. Process failure mode and effect analyses.
2. Process control plans with designated reactions to out of specification conditions.

3. Designated inspection frequency, often 100% (test stands).
4. Process capability requirements and application of SPC.
5. Types of documentation are specified, especially for the safety category.

More recently, critical characteristics have been used to identify characteristics critical to fit. Since most complex assemblies are analyzed by statistical methods, critical characteristics for the fit category allows dimensions used in statistical stack-ups to be identified and the appropriate controls applied when their tolerances have been identified as significant contributing factors. Two symbols are suggested in Table 5-1. The first identifies the more frequent case where the tolerance is bilateral and process centering is required. The second symbol identifies processes where bias is desired, usually where a unilateral tolerance is shown.

For many dimensions, the most important factor, and one which manufacturing personnel can most readily control, is *process centering*. Conventional quality control methods using C_p and C_{pk} parameters do not directly foster process centering. In fact, for a given tolerance, C_{pk} parameters allow greater centering error as the process spread diminishes. Where dimensional bias is desired, as denoted by the use of unilateral tolerances, the conventional quality control wisdom of striving to increase C_{pk} values yields the wrong results!

To support characteristics which require centering, a new parameter, C_c, was described in Chapter 4 where the author suggested compliance limits of $C_c \le 0.25$, $C_p \ge 2.0$,

Table 5-1.

Symbol:	Category:		Compliance Requires:	Continuing Improvement:	Applies to:
▽ Or ⦿▽	Safety & Compliance		1. Strict (100%) compliance to specification 2. Documentation (traceability)	Increase C_p Document	Features and specifications that affect safety or compliance.
◇	Function		Statistical compliance to specification	Increase C_p	Features and specifications that affect function, eg. time to failure.
◇ (C)	Fit	Bilateral tolerance	Center Process $C_c = \frac{\overline{X}}{T} \le \pm .25$ $C_{pk} \ge 1.5$ $C_p \ge 2.0$	Reduce C_c Increase $C_p - C_{pk}$	Any dimensions used in a statistical calculation to meet a vehicle goal.
◇ (B)		Unilateral tolerance	Bias Process $C_{pk} \ge 1.5$ $C_p \ge 2.0$	Increase C_p Optimize $C_{pk} = 1.5$ (nominal side)	

and $C_{pk} \geq 1.5$. These values are consistent with the 6 sigma process control concept. Continuing improvement efforts attempt to reduce the C_c value to zero, while never allowing it to exceed 0.25. Another way to achieve the $C_c = 0$ condition is to drive the C_{pk} value to equal the C_p value as shown in Figure 5-10a.

Figure 5-10 provides a perspective to the cycle of constant improvement. As previously noted, Figure 5-10a is focused toward centering a process and is appropriately used with bilateral tolerances. Figure 5-10b is focused toward unilateral tolerance applications and shows that the process should be biased to obtain an optimum C_{pk} value of 1.5 (or 1.0 as the case may be) at the nominal side of the tolerance range. This application is often overlooked in standards that stress increasing C_{pk} values for all applications.

A final word on process improvement: Basically, it is an economic decision and priority should be given to critical characteristics and other toleranced features where a loss can be demonstrated for extreme tolerance occurrences.

a) Bilateral tolerance (±T)

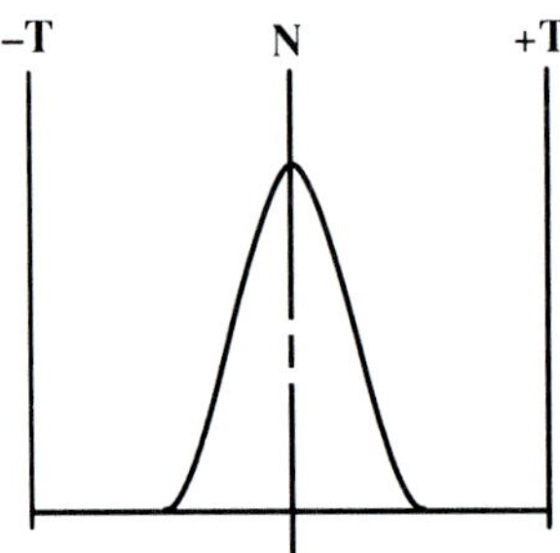

C_p increase
$C_{pk} = C_p$ (centered)

b) Unilateral tolerance $\left(\begin{matrix}+T \\ -0\end{matrix}\right)$

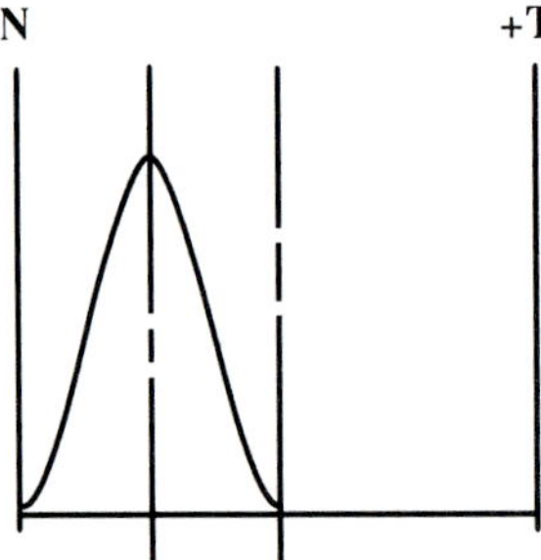

C_p increase
$C_{pk} = 1.0$ (nominal side)

Figure 5-10. Process Improvement

Figure 5-11 illustrates that a focus only on the C_{pk} index can fail to properly target process means. Here mating parts with high C_p and reasonable C_{pk} values resulted in an assembly problem as mating parts were effectively at extreme tolerance limits.

Figure 5-12 illustrates an automobile decklid to fender gap and the controls applied to the decklid to meet vehicle goals. This part was chosen for illustration purposes as it displays both of the fit classes: centered and biased.

Finally, it will become apparent in the use of critical characteristics, that each designated product characteristic will result in one or more process characteristics. Manufacturing and quality controls are actually placed upon the process characteristics. Uniform symbols for designating process control have not been developed, however, some companies

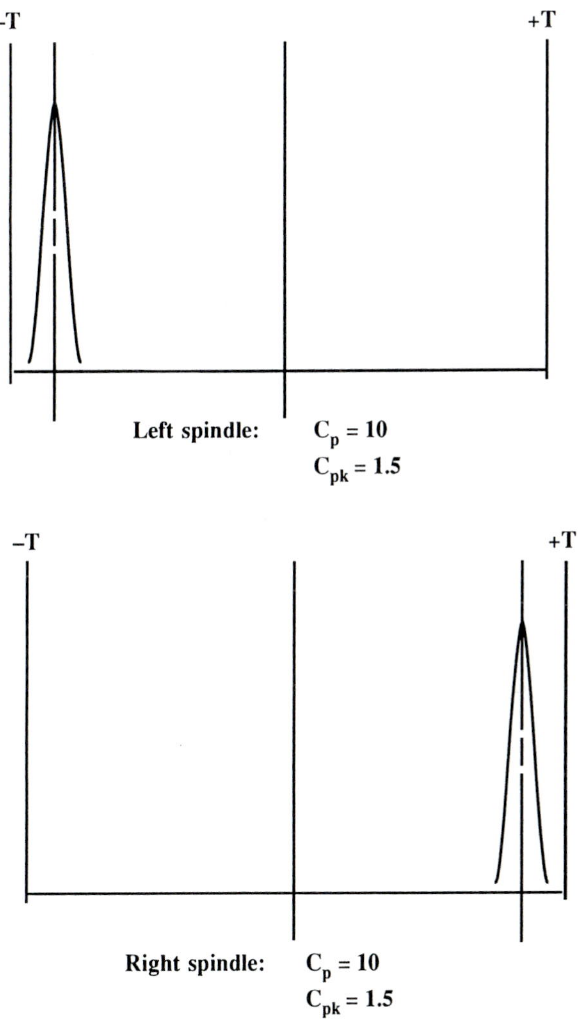

Figure 5-11. Centering Problem

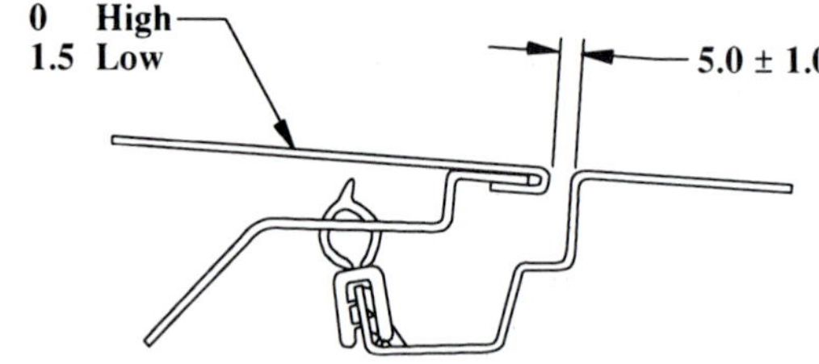

Section 38 goals

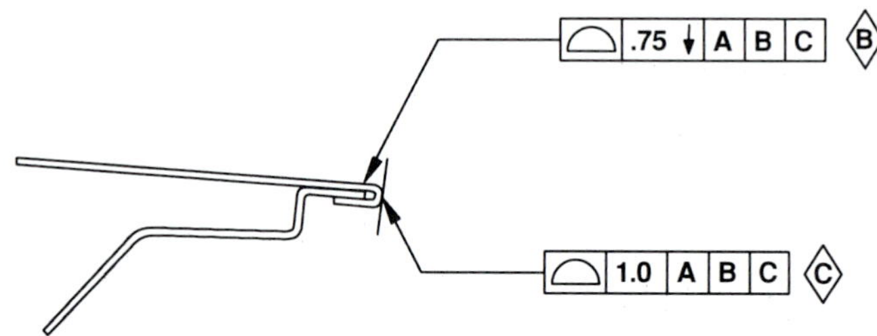

Figure 5-12. Decklid controls

add sequence numbers to their product characteristics to cross reference the critical characteristics found on the product drawing with the corresponding process characteristics found in a process control plan.

SUMMARY

GD&T is a symbolized tolerance system that has the unique ability to convey product function.

Symbolized tolerance notes are desirable because they:

1. Increase clarity of meaning.
2. Reduce documentation time.

3. Are language independent.

Symbols have been developed to describe:

1. Types of tolerance (e.g., position tolerance).
2. Projected and fastener heights.
3. Critical sizes.
4. Datums.

Critical sizes indicate the feature size at which the datum or tolerance requirement applies. Critical size categories include:

1. Maximum material condition (MMC).
2. Least material condition (LMC).
3. Regardless of feature size (RFS).
4. Virtual condition (VC).

Critical characteristics communicate the engineer's knowledge downstream and identify features which require strict (100%) or statistical performance to specifications. Critical characteristics also can identify dimensions and tolerances used in statistical stack-ups so that appropriate controls may be applied.

Chapter 6

Datums

Datums are the origins from which locations are measured and the reference system for orientation controls. They are used with geometric and position tolerances to define clearly the surfaces or features that position and constrain the toleranced component so that specific tolerance requirements can be measured from theoretically perfect datum planes that are represented by precise manufacturing and checking equipment. Machine tables and surface plates are not true planes, but are usually of such high quality that they simulate theoretical datum planes adequately. Measurements, then, are made from planes and axes in the processing equipment.

COMPONENT POSITIONING

Everyone is familiar with one form or other of a component which is securely fastened into an assembly. Few of us, however, have given much thought to the means which are necessary to position any component in a secure manner. Figure 6-1 illustrates that for any three-dimensional system, six degrees of freedom exist. In other words, movement may occur in the direction of the three independent axes, X, Y, and Z, and rotation may occur about any of these axes. To secure a component against all possible motion and to establish a unique position for the component, it is necessary to resist each of the six degrees of freedom. Stated in another manner, it is necessary to resist movement in three directions and to resist rotation about three axes.

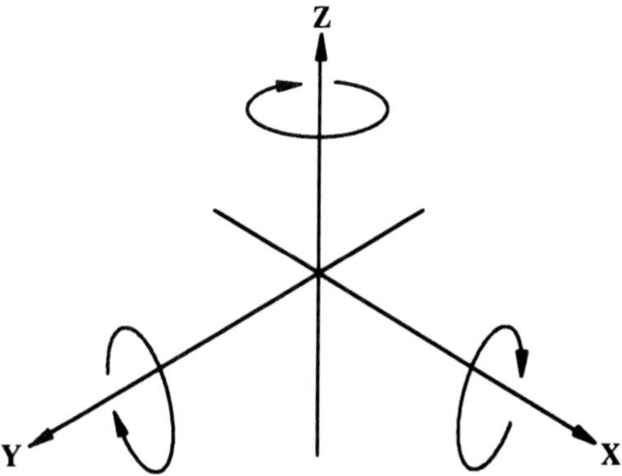

Figure 6-1. Illustration of six degrees of freedom

METHODS OF CONSTRAINT

Three-Plane Datum System

The axis system of Figure 6-1 can be represented by three datum planes as shown in Figure 6-2. For many regular-shaped parts, particularly those with flat surfaces, the three datum planes will position the part and constrain it against movement and rotation. Because flat machined parts occur so frequently, the three datum plane system was one of the earliest defined specification systems. In this system, depending upon the tolerance requirement, up to three datum planes are designated to orient a component, the order being extremely important due to possible imperfect geometry of the component being datumed.

The datum listed first is the primary datum plane, and the corresponding surface of the component will contact the *XY* datum plane at three or more points if the surface is irregular and a stable position exists, restricting motion in the *Z* direction and rotation in the *XZ* and *YZ* planes.

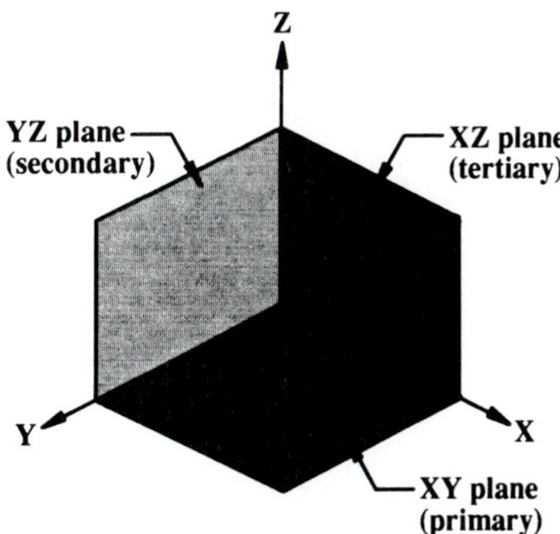

Figure 6-2. Three plane system

The second surface specified contacts the *YZ* plane at two or more points, again depending upon the quality of the surface, and restricts motion in the *X* direction and rotation in the *XY* plane.

The third surface will contact the *XZ* datum plane at one or more points, providing the final constraint against motion in the *Y* direction.

In each case, the datum plane is not the actual surface of the part, but rather a theoretical plane with which the specified datum surface comes into contact. In practice these theoretically perfect surfaces are represented by surface plates, fixtures, and other surfaces with a relatively high degree of accuracy. Figure 6-3b shows the sequence of positioning for the component and the datum system specified in Figure 6-3a.

The three-plane datum system assumes that the surfaces of the component considered have a high degree of squareness and flatness. These are necessary conditions because an improper number of contacting points caused by out-of-squareness, bowing, warping, or

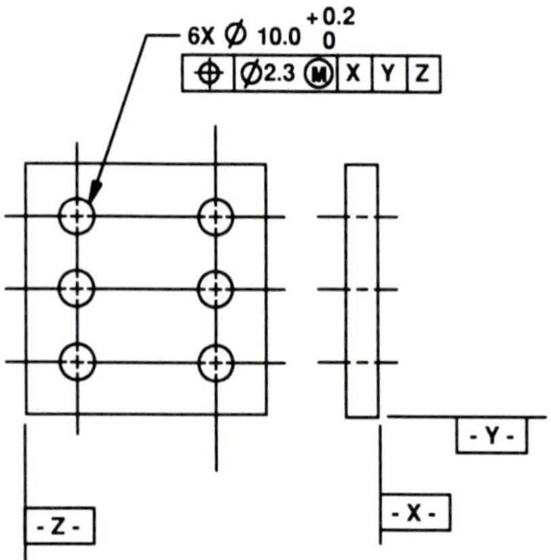

a) Component with a functional datum

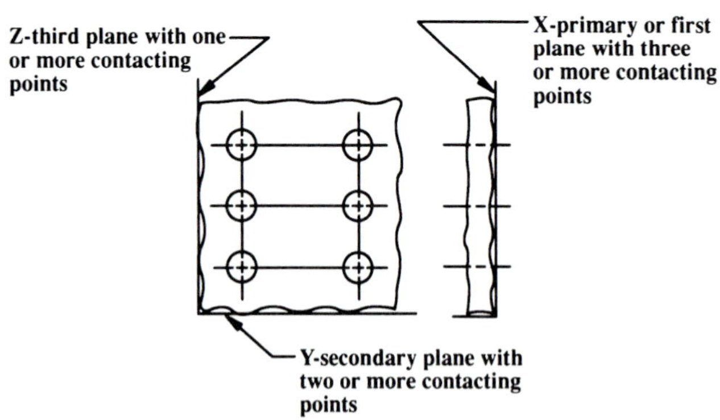

b) Interpretation of the datum specification

Figure 6-3. Example of 3 plane system

surface irregularities and, in extreme cases, the part may be distorted or may actually rock. If a part is too irregular, or too flexible to be stable and capable of being located repeatedly against the three planes, other techniques, which follow, must be used.

In some cases, the designation of a plane is not only a good description of product function, but is also reasonable in that the component surface has a high degree of flatness. The broached surface of an internal combustion engine or a compressor head would be a good example of such a part. Other parts, most often forging, casting, heavy metal stamping, and weldments, do not have sufficiently accurate surfaces to utilize this system. Flexible parts, including some light gage metal stampings, and many non-metallic components, such as plastic and elastomeric parts lack the rigidity to use this system without further refinement.

Datum Target System

Figure 6-4 illustrates the general method of constraining a part against six degrees of freedom through the use of six contacting points. Points 1 and 2 contact the *XY* plane and prevent motion along the *Z* axis and rotation about *Y*. The addition of point 3 prevents rotation about the *X* axis. With only these restraints, however, the component would be free to turn about the *Z* axis or slide along the *XY* plane. Point 4, contacting the *XZ* plane, limits motion along the *Y* axis, while its companion point, point 5, prevents rotation about *Z*. Point 6 prevents motion of the part along the *X* axis and thereby uniquely defining its position.

Castings, forgings, weldments, and other irregularly shaped rigid parts are often positioned by six contact points called *targets*. Such target points are often added to irregular shapes specifically to aid in positioning the part during machining and inspection operations. On less complicated parts target point specifications are often used to identify points used in manufacturing operations, thereby eliminating variations in the production of the part which might occur from positioning an inaccurate part against a full plane. The obvious alternative to such a practice would be to impose additional controls, such as flatness, perpendicularity, or parallelism, upon the locating surface. Where such controls are not required to maintain part function, the extra operations may be unduly costly and, therefore, are often impractical.

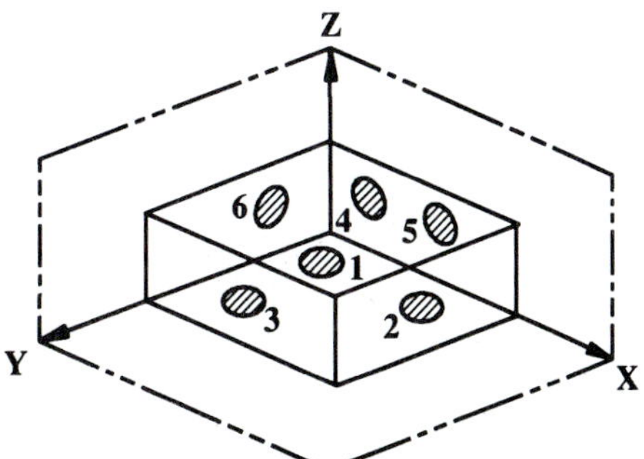

Figure 6-4. Component constrained against 6 degrees of freedom by targets

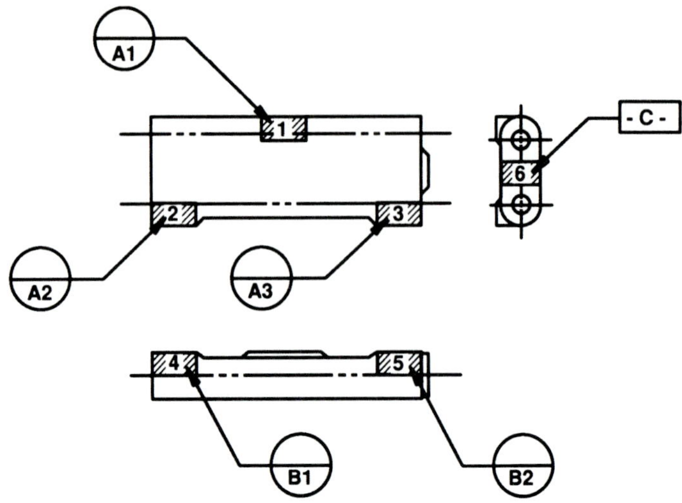

Figure 6-5. Typical specification

Figure 6-5 illustrates the use of target points on an irregular shaped casting. General practice is to group the points in common planes and use three datum plane notation. In practice, you frequently can experience parts where all the points do not lie in a common plane (e.g., point 1 located at a different level than points 2 and 3). Although a composite datum designation might be used, it is a common practice to denote primary and secondary target groups as though they were in true planes. Figure 6-12 is an example of datum targets not contained in simple flat planes.

Other Datum Systems

While it is possible to rationalize complex sets of features into planes, the fact remains that datums are specified as features and sets of features and, in many cases, they do not readily appear to constitute three planes. Viewing a datum set as a group of features, many unique datum systems can be experienced. The fundamental observation to be made in such cases is whether or not the part is sufficiently constrained against the possible six degrees of freedom to permit the specified tolerance to be measured. Examples of such datum systems are shown in Figures 6-6 through 6-8.

Figure 6-6 displays a datum requirement where one datum is the primary plane and the second is the *B* axis. Since the part has no other features than the equally spaced holes, it is not necessary to constrain rotation about *B*. If a feature such as a keyway was present at the part outside diameter, a third datum could be added to constrain the possible rotation about *B*.

Figure 6-7 shows a composite datum consisting of an axis through features *A* and *B*. The *A-B* designation indicates one datum as opposed to two different levels of priority. In this case, the datum limits only four of the degrees of freedom.

Figure 6-8 shows a common datum system of a primary surface (*A*), an axis (*B*), and the final constraint against rotation (*C*).

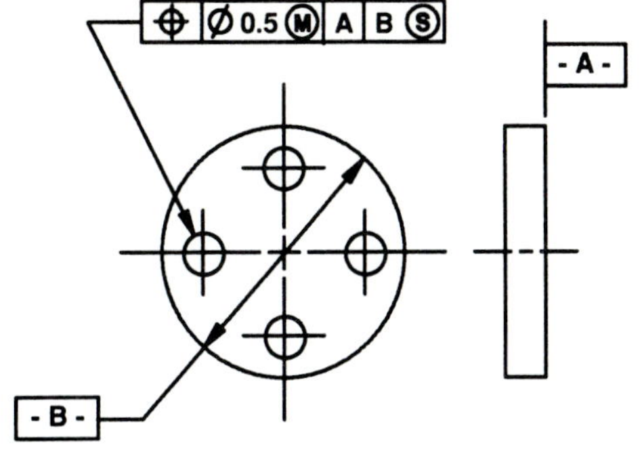

a) Specification

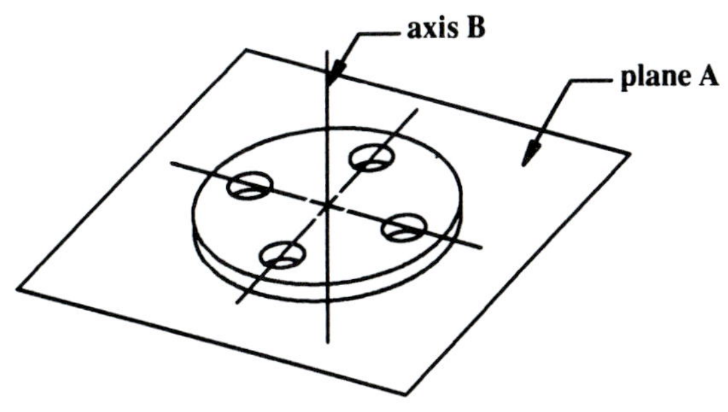

b) Interpretation

Figure 6-6. Datum plane and axis

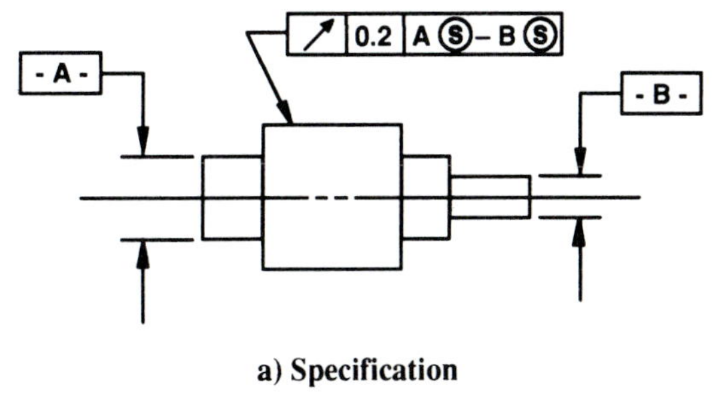

a) Specification

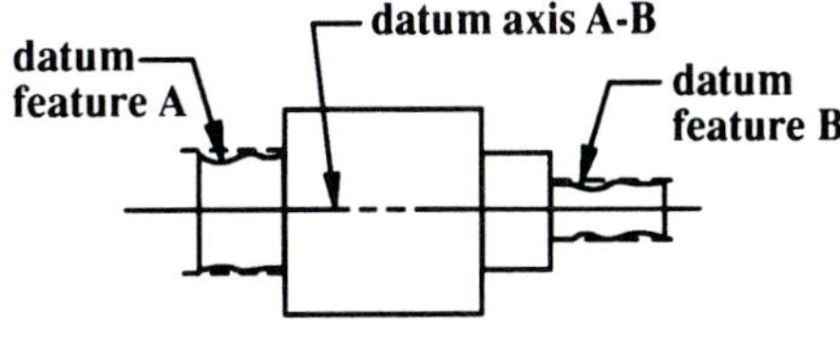

b) Interpretation

Figure 6-7. Datum axis

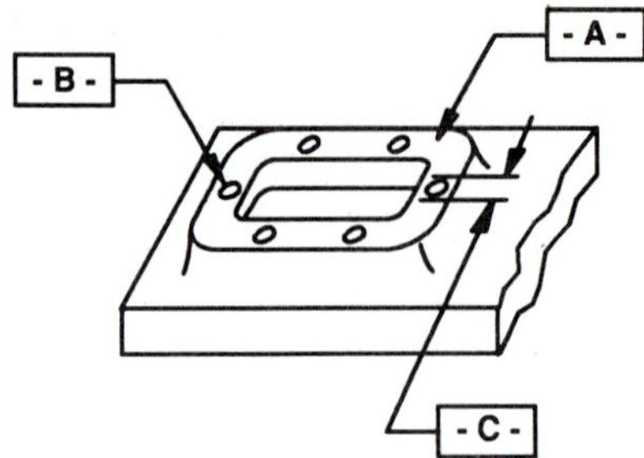

Figure 6-8. Datum system comprised of one plane, one axis, and anti-rotation feature

NON-RIGID PARTS

There are many types of parts that can be classified as *non-rigid.* The non-rigid classification includes many levels of flexibility from parts like glass or sheet metal which are flexible normal to their primary plane, but rigid within that plane. Other parts, such as wire forms, molded plastic, and so on are flexible in several directions. Examples of non-rigid parts include:

1. Flat panels of most materials.
2. Formed thin metal parts.
3. Plastic and other material with low modulus of elasticity.
4. Parts with extensive length versus cross section, such as tubes or formed wires.

Non-rigid parts may use any of the datum systems previously described, however, the issue of restraint must also be addressed to ensure meaningful and repeatable measurements. If a flexible part is restrained to its datum system, it may be considered as a rigid part. The applied restraint must simulate the mating part function! In most cases, but not all, the restraint will be limited to the primary datum surface. Figure 6-9 typifies a sheet metal part that is restrained at 6 points to create a primary datum surface *A*. Datums *B* and *C* are a hole and surface respectively. The datum targets are numbered *A*1 through *A*6 indicating one datum plane. The target symbols denote a 32mm square contact area and an applied force of 40N to true the part. If the force is inadequate to fully flatten the part a higher force should be considered unless the 40N represented a functional limit, such as the force an attaching bolt could apply, or the limit of clamping force in an assembly operation. Caution should always be used when increasing restraining force as many other problems can result from stressed components.

The typical specification shows primary datum, surface *A* restrained to 6 points. The restraint is noted by the symbol®[1] and the force limits at the target points. Hole *B* is noted

[1]Recommended by this author and not yet included in any published standard.

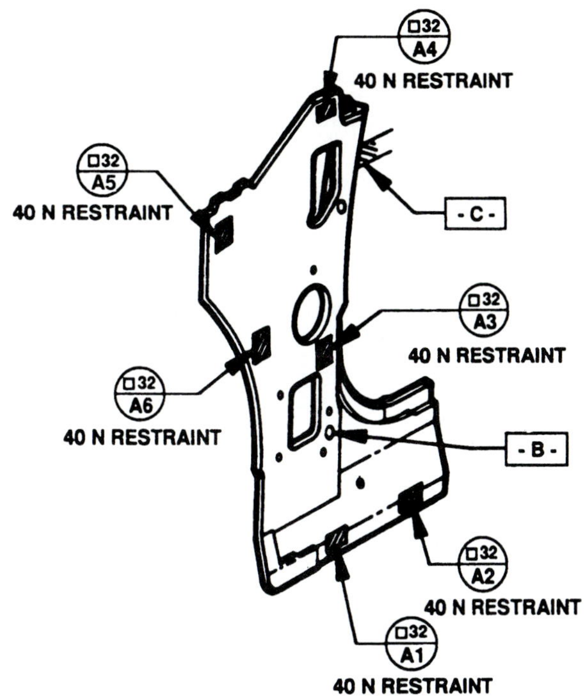

Typical tolerance specification

Figure 6-9. Restrained non-rigid part

to apply at a critical size of maximum material condition, denoted Ⓜ. Surface *C* provides the final datum feature.

An alternate method shown in Figure 6-10, is to specify 3 target points as primary datums, say points *A*1, *A*2, and *A*3. The other points are checked for error in the free state and then clamped to form the full datum plane. For the previous example points *A*3, *A*5, and *A*6 could be designated *D*1, *D*2, and *D*3. The datum for the final checks would then be a composite datum, *A-D*. Thus, distortion can be managed with limits on either restraining force or on dimensional variation. Force level may be best in most cases because it is easier to rationalize a force level limit than to correlate dimensional variation to assembly distortion.

Supporting Points

Occasionally, you'll see a specification that requires free state points to be supported prior to measurement. This does not apply a datum relationship, but rather is a simple recognition of the flexibility of the part. The intention is to back up (but not locate) a point that could be deflected in the measuring process, thereby contaminating the actual measurement.

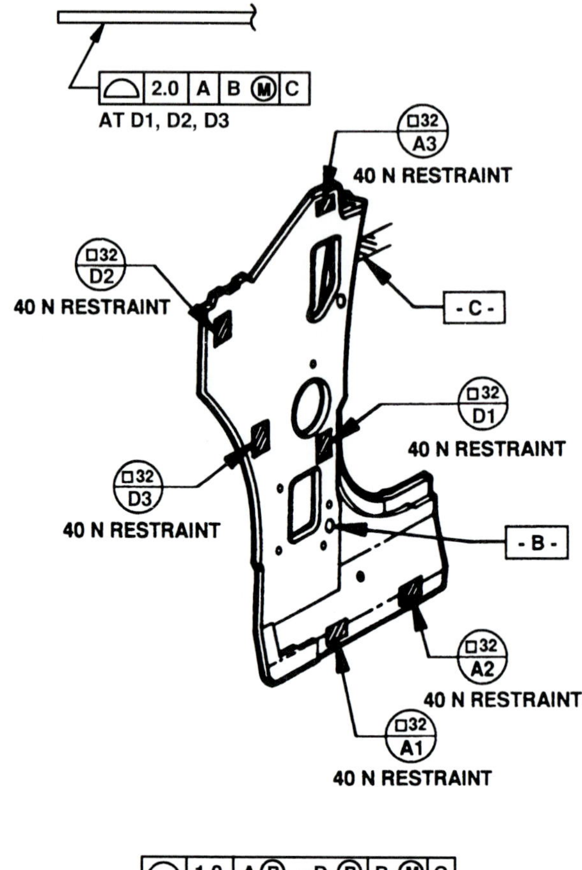

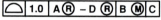

Typical tolerance specification

Figure 6-10. Non-rigid part with free state check

DATUM APPLICATION

When components are toleranced to control their variability, the first step is to select the datums. If the part is simple, the datum selection is equally simple. As part complexity increases, datum selection often becomes more difficult. Complex parts often have many possible choices of datums. As shown in Figure 6-13, the choice of datums on such complex parts will affect how the part variation is distributed when it is assembled.

Datums used in tolerance specifications can be classified as functional, convenient, or special datums. Each of these types can be employed to fit specific needs. In all cases, however, datum features must be accurate and easily identifiable; must be accessible on the part; and must be of sufficient size to enable manufacturing and inspection operations to be preformed. Once established, datums must be used as specified for all manufacturing and inspection operations.

Datum Coordination

Datum coordination is a term that means different things to different people. Since the term seems to imply that it is a requirement, narrowing the definition seems appropriate. The most appropriate application of the term is to a part which passes through several stages of subassembly, as in the case of a component used in a welded structure. If the datum targets used for the component are shifted when it has parts or operations added, small shifts may occur due to form errors in the part as in Figure 6-11a. Here moving the datums on a warped flange will move the part in an up-down direction. As shown in Figure 6-11b, a full surface datum avoids the interaction that can occur with datum targets and form errors. Hence, when appropriate, the use of full surface datums avoids coordination issues.

Aligning datums on adjacent parts, as in Figure 6-12 is not true coordination, but is often preferred although it does not in itself reduce variation. Usually the datum pattern is repeated due to its sound application (e.g., widespread for stability). Hence, slightly misaligned datums on adjacent parts is more a problem with logic and comprehension than with variation management.

Datum Sequence

As noted in the preceding chapter, the *datum sequence* in the feature control frame specifies the datum sequence in manufacturing and inspection operations. The specified sequence of datums should be based upon the functional requirements of the feature being toleranced and should simulate the manner in which parts are fitted together in its assembly. The implication of this approach is that the datum sequence may be different for different features.

While this may be satisfactory for simple parts with few GD&T requirements, or for

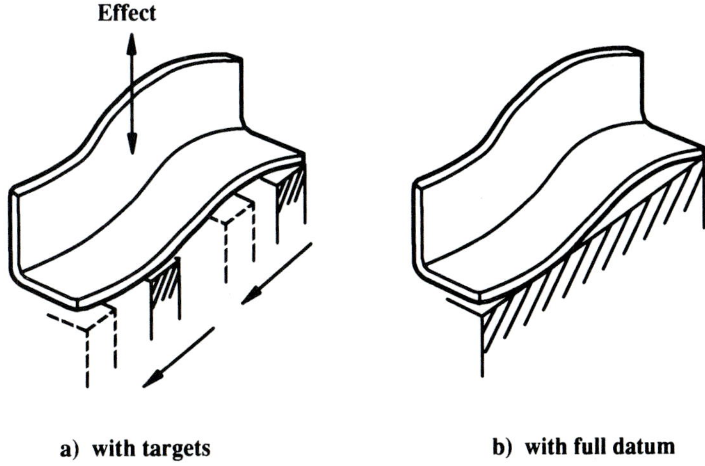

Figure 6-11. Datum coordination

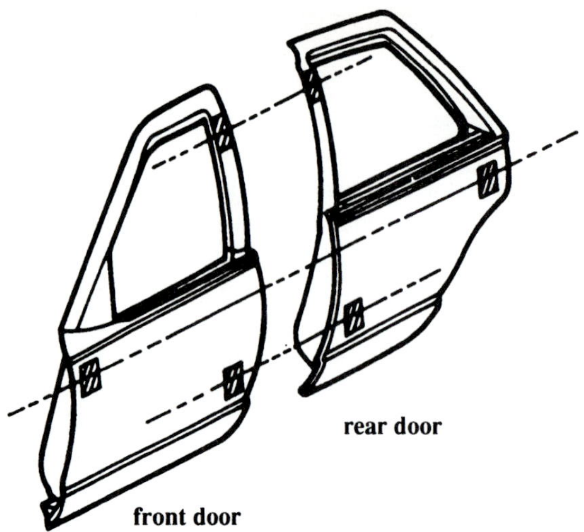

Figure 6-12. Datum coordination

machined parts with multiple functional interfaces, complex parts with many requirements, such as curved sheet metal parts, require another approach. Here a principal datum system is selected and used extensively throughout the part for *location and attitude tolerances*. In this case, the specified principal datum system is based upon the functional requirements of the part being toleranced and simulates the manner in which the part is fitted into its next assembly. For example, the principal datum set *A*, *B*, *C* in Figure 6-15 represents the mounting and alignment points of the fender when it is installed to the vehicle body. Although many datum sets seem possible on such complex parts, each part is fitted into its next assembly in a defined sequence and study of the process and its assembly goals will yield an appropriate datum set. A principal datum system also limits the number of inspection setups required.

Functional Datums

Functional datums are the highest level datums and should be used whenever possible. To recognize a functional datum requires that the analyst understand the function of the component. That function may be a design feature, such as a pilot that locates a pattern of holes or the interface of mating parts. Knowledge of the process may also be required for complex parts. For example, an automobile door may use locators (datums) on the outboard panel to locate the door while the hinges are attached. Here the function is the desire that the door panel be flush to other body panels. Hence, the process sequence can affect the datum selection.

Complex parts with many functions can have many possible datums. For example, in Figure 6-13 the effect of datum placement on door gap is summarized. On this vehicle the hinge is attached to the door and then pierced to create an alignment hole that matches an alignment hole in the body. Hence, datum placement in the piercing operation affects the

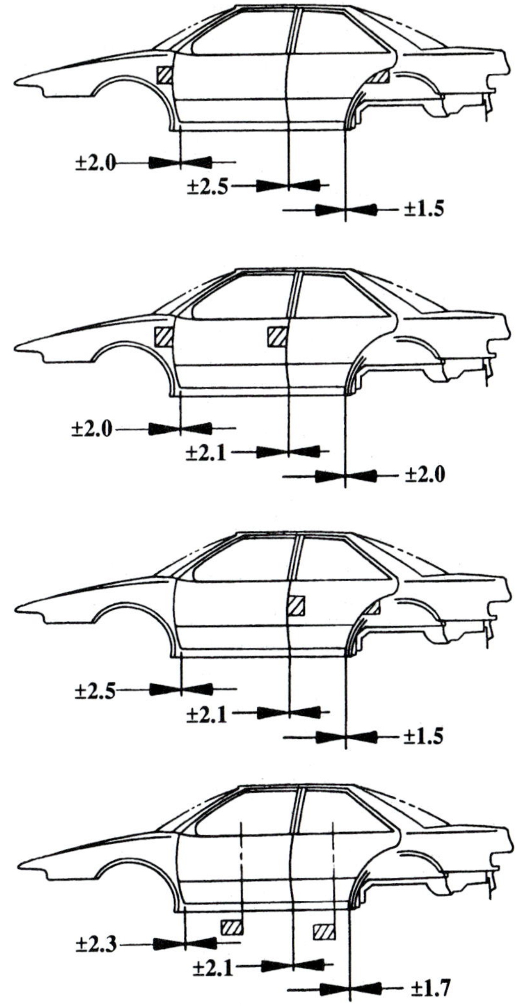

Figure 6-13. Alternate datums

variation in door edge gaps as the door width tolerance is large compared to the position tolerance for the pierced hole. The datum choices for the two holes were:

Option	*Front Door*	*Rear Door*
a	Front Edge	Rear Edge
b	Front Edge	Front Edge
c	Rear Edge	Rear Edge
d	Center	Center

Although it had been anticipated that the best control would be obtained by option d, option b proved to have the best balance of error distribution. In retrospect, the result was easy to understand. Figure 6-14 shows the controls applied to the mating points on the body. Here the front fender mounting holes are located from the rear quarter panels, datum *B*. Hence, in Figure 6-13 the fender errors influence the forward gap, while the rear gap, being a datum surface is relatively error free. In option b, the front door width errors influence the center gap and the rear door errors influence the rear gap. Hence, each gap is influenced by one major source of variation and balance in the expected variation results.

Figure 6-14 also serves as an example of a complex datum system. It represents an operation that pierces and forms critical attaching points on a completed body. The machine computes datum *B* from the average of the two sides, hence, the unusual datum specification *B average* and the reference note that tells how the body is positioned when measurements that establish *A* and *B* are made.

Auxiliary Datums

Auxiliary seems to be common, but misunderstood, datum terminology. In the past, a datum used to constrain a part by limiting rotation about a datum axis (e.g., a keyway) was called auxiliary. Clearly, any datum needed to constrain a part against the six degrees of freedom is part of the basic datum structure and is not auxiliary. The best current use of the term relates to a second functional datum system placed upon a component to control tolerances in a local area. This is frequently found with parts that use a principal datum system for a

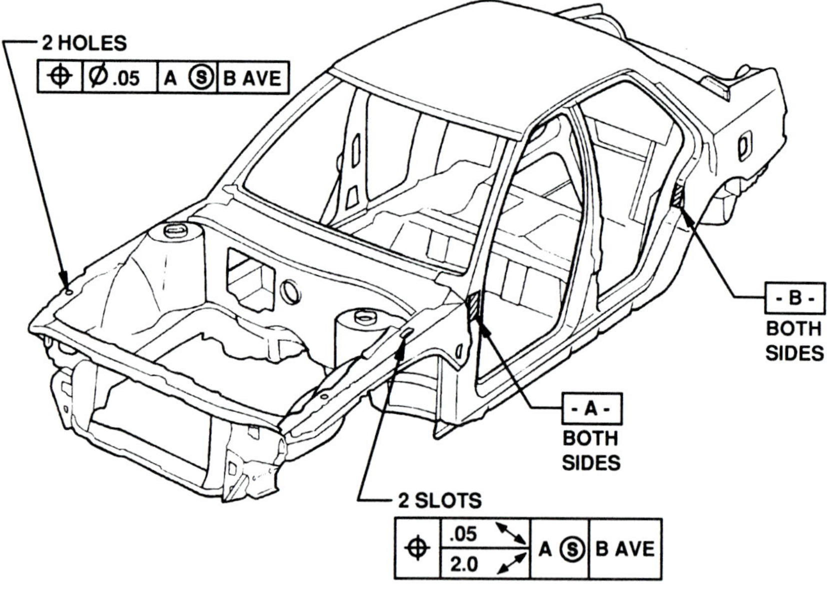

Note: All of the G.D. & T. requirements included on this specification assume the body in white is positioned upon its basic car track datum support.

Figure 6-14. Complex datums

majority of the functional controls. A tolerance between the principal datum system and the auxiliary datum will be appropriate in most cases. In Figure 6-15, the basic datum system is *A*, *B*, *C* and auxiliary datum system *J*, *K*, A_4 is used to locate the mounting holes for the front corner light.

Convenient Datums

A *convenient datum* is a feature selected for use in manufacturing and inspection operations when there is no functional datum or where a functional datum proves unusable (e.g., one which is not sufficiently stable). Restriction of the size and location tolerance of one or more holes within a pattern for manufacturing purposes is a common example of the convenient datum. The convenient datum may be designated on the product drawing or may be introduced in a manufacturing specification. Current trends would include it on the product drawing following consultation with the manufacturing engineer involved.

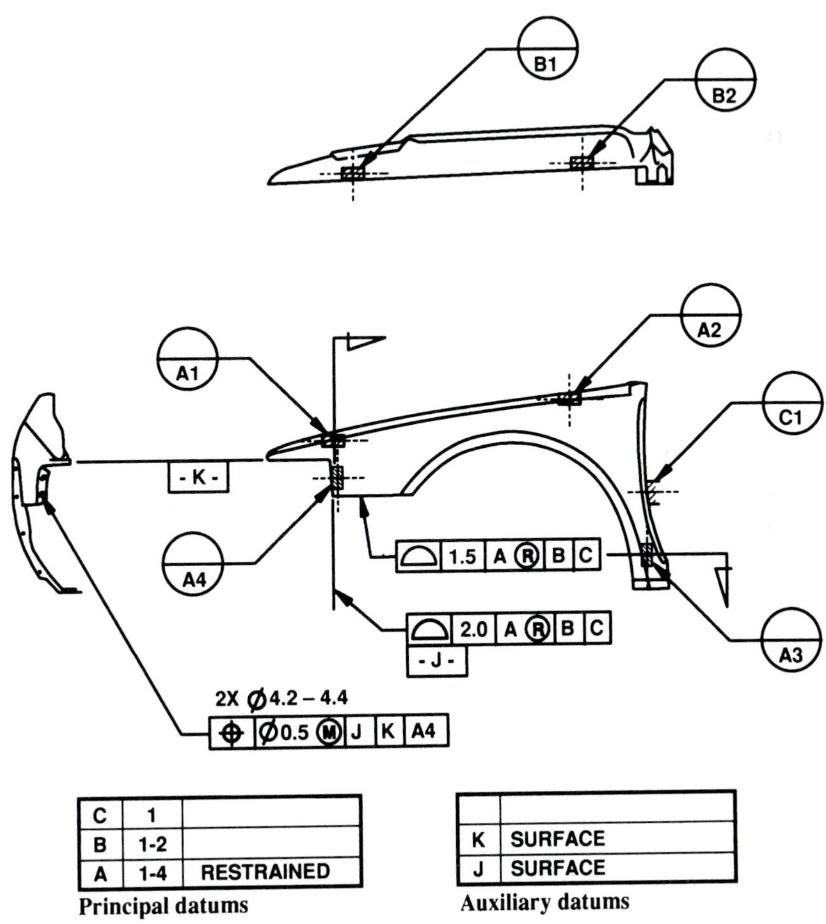

C	1	
B	1-2	
A	1-4	RESTRAINED

Principal datums

K	SURFACE
J	SURFACE

Auxiliary datums

Figure 6-15. Component with principal and auxiliary datums

Special Datums

The *special datum* is similar to the convenient datum except that it is a feature (such as a hole for a fixture location pin) added to a component expressly for use as a tooling origin. It is usually selected because of its advantageous size or location. Such datums are usually requested by manufacturing departments, but are often specified on the product drawing since they require product engineering review of their impact upon part function. This also avoids apparent discrepancies which could arise in the inspection of such components should features be found which are not shown on the product drawing. *Gage holes* used in stamped metal parts to position them in welding operations are good examples of special datums.

Implied Datums

Many corporations have not adopted the use of datum specifications and, in those corporations where GD&T is used, there may be hundreds of active drawings which were made prior to its adoption. Therefore, it is very likely that situations will be encountered in which a feature, although not specified on the product drawing, is a datum surface due to its functional nature or use in a critical manufacturing operation. Such a surface is called an *implied datum.* Gages, fixtures, and machining operations must maintain the functional requirements of implied datum relations, as well as those explicitly stated.

The interface of mating parts is an example of the implied datum. Such an interface must be held to be functionally important because position tolerance limits, for example, are by definition constructed normal to the interface of mating parts. In many cases (e.g., a part with tapped holes), the interface may be easily identified. With other parts, such as those constructed of flat plate, the choice may be less obvious and require more detailed study.

When one has the task of updating datum requirements on existing parts, both the product function and the existing fixtures and gages should be studied to determine both desirable and actual datums which can then be translated into appropriate GD&T.

CRITICAL SIZE

All datum features of size are subject to *critical size* relations. Critical size relationships are discussed in more detail in Chapter 9. Datums may have any of the three possible types of critical size although in practice LMC relations ships are seldom found.

RFS Datums

Functional datum features which assemble with an interference fit result in the *RFS type* of datum. This type of application, illustrated in Figure 6-16, also applies to transition fits in which either small amounts of clearance or interference may exist at the assembled datums. Examination of this assembly leads to the conclusion that the pilot (Diameter *A*) will forcibly position the fastener holes. With an interference fit, variation in the size of the pilot and the pilot bore will have no effect upon the hole locations because a fixed datum axis is

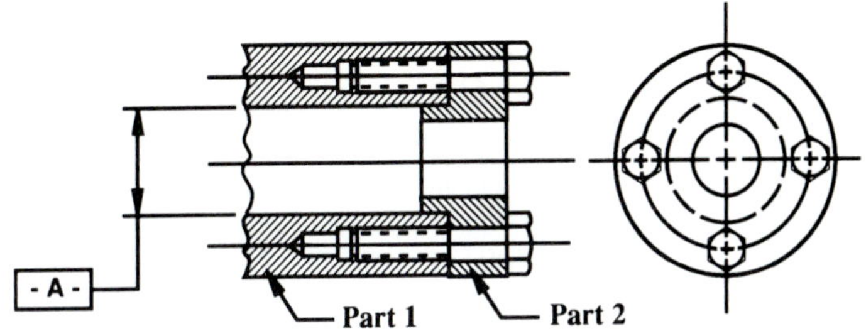

Figure 6-16. Assembly with datum features which assemble with an interference fit

established for both parts 1 and 2. Since the axis of the pilot surfaces are the datum points for determining the true positions of the hole pattern, the datum surfaces are said to be critical regardless of feature (datum) size, symbolized by Ⓢ. Specifications of this type necessitate inspection of the feature pattern from the axis (or center plane) of the datum feature. For example, the part illustrated by Figure 6-17 would be inspected by determining the measurements from a plane perpendicular to surface A through the exact center of diameter *B* and keyway *C*.

Special datums, such as gage holes, are frequently specified as RFS datums. Automotive body stampings sometimes use gage holes in detail parts as principal datums to control the myriad of features on the part. When such parts are assembled, they are positioned by the gage holes to transfer the relations from the detail to the assembly level. As such, it is appropriate to designate the gage holes as RFS so as to avoid any accumulation of tolerance when checking at the detail level and locating at the assembly level. Practically, the RFS designation denotes the use of tapered or expanding pin locators to achieve the RFS control in fixtures and gages.

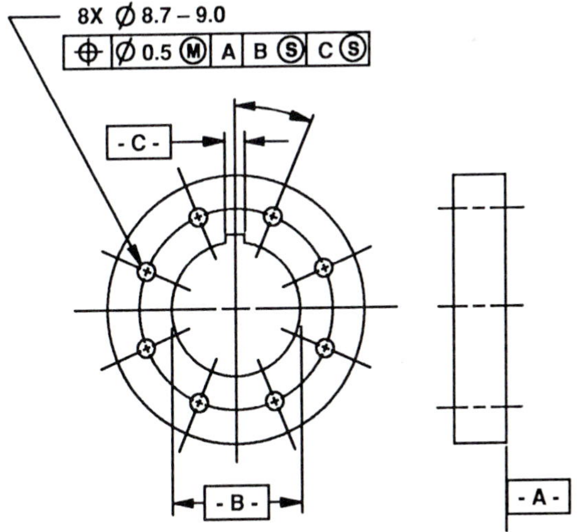

Figure 6-17. RFS datum specification

MMC Datums

The second type of functional datum is associated with datum features which assemble with clearance, often approaching zero at their *maximum material condition* (MMC). Referring again to Figure 6-16, it is obvious that when assembled datum surfaces are at MMC, and clearance is zero or very small, the relationship is the same as the RFS relationship. When the pilot diameters are not at MMC, the clearance which exists will allow movement of the datum surfaces relative to the hole pattern. In this situation, there is an allowable datum tolerance (or shift) which is proportional to the clearance between assembled datum surfaces, which in turn is proportional to the size deviation of each datum surface from its maximum material condition. Because the datum tolerance varies with size (approaching zero at datum maximum material condition), such datums are said to be critical at maximum material condition.

Further consideration of MMC datums indicates that the value of datum tolerance available follows the formulas for additional tolerance developed in Chapter 9 since the critical tolerance is equal to zero. Formula 9-2, developed for MMC clearance holes, obviously applies to internal datum features, while Formula 9-4, developed to describe MMC projections, describes the external datum feature equally well. Datum tolerance zones, like all position tolerance zones, are three-dimensional in nature. The height of the tolerance zone is equal to the height of the engaged, mating datum surfaces. Datum depths should be clearly designated as in Figure 6-18. Practically, MMC datums in fixtures or gages are pins (for holes) or holes (for projections) sized to the MMC limit of the toleranced feature. MMC datums are popular from the viewpoint of manufacturing personnel as the pins and holes are simple and reliable. Also, since clearance usually exists, such locators do not *hang-up* as do RFS locators, and hence, are less troublesome in use.

Virtual Size Datums

When two or more MMC features are specified as a datum, they must be considered as *virtual size datums*. Figure 6-19 illustrates such a case. Here, two holes marked *X* are the datum for two additional holes. When you consider use of the two *X* holes for datum pur-

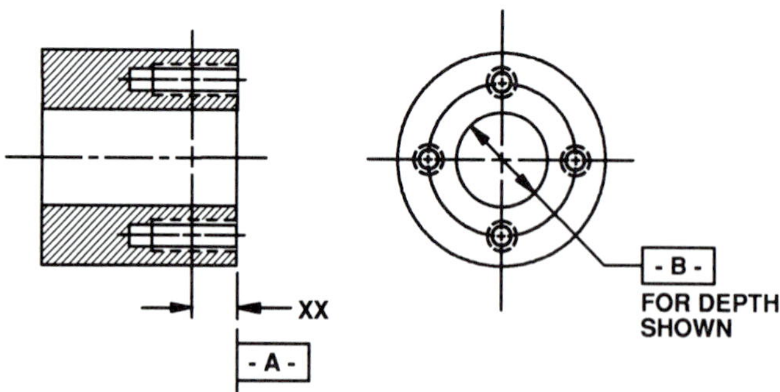

Figure 6-18. Datum depth

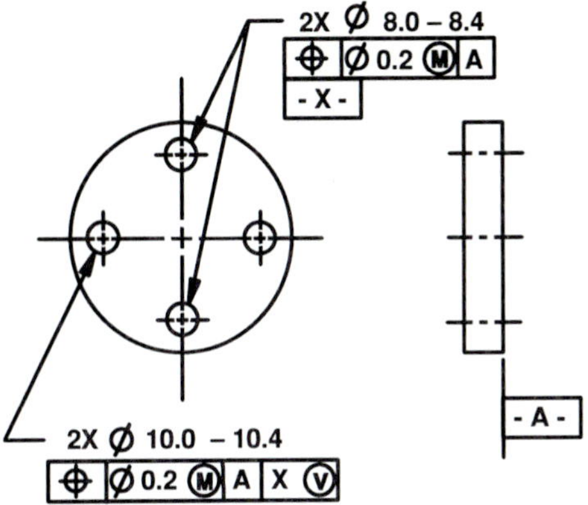

Figure 6-19. Virtual size datum

poses, their MMC limit is the obvious starting point. However, since these holes also are affected by a ∅ 0.2 position tolerance, that tolerance must also be considered when using these holes as datums. Such features are designated Ⓥ for virtual size. Actual values of the virtual size may be calculated from Table 12-1. The same results would be achieved if appropriate notes were applied to indicate that the 4 holes constituted one pattern.

Datum Specification

Datum specifications follow the rules established for *feature control frames.* Two specification issues surface relative to the quality of the specification:

1. A sufficient number of datum planes, axes, and/or features must be denoted to constrain the part against 6 degrees of freedom. If the tolerance does not require restraint in all 6 degrees, a lesser number may be used, as in Figure 6-7. Complex parts with a principal datum system often repeat the full datum requirements for all tolerances, even when all the datums are not necessary on some tolerances. Generally, this is not a problem. However, where profile tolerances are involved, caution is recommended as the only easy way to distinguish *profile form tolerance* from *profile location tolerance* is the absence of datums. As such, the answer to how many datums should be included in the feature frame is *enough to satisfy the function.*

2. The critical size modifiers Ⓜ or Ⓢ or the Virtual Size symbol Ⓥ should be included on all datum features of size. Current standards infer a critical size for datums depending upon the type tolerance specified. This author feels that the designations should always be included because they are too important to be misunderstood. Further, use of the implied datum systems can result in a principal datum system being denoted in two ways on a common part, depending upon the type of tolerances used. This is too troubling to consider as a competent method of specification.

DATUM TOLERANCE

There are several ways of allowing tolerance between a datum and features that are related to the datum, and, as one might expect, several ways of expressing the tolerance control.

Datum Tolerance

Datum tolerance or pattern location tolerance are two ways of saying that a datum may move relative to other features, such as a pattern of holes, within some specified limit. Refer to Figure 6-20, an assembly similar to Figure 6-16 but with clearance between the two pilot (datum) at all times. It is obvious that the clearance can be used as a datum for a tolerance. Such datums can be seen to be simple *fixed fastener applications.* The critical tolerances are calculated according to the formulas given in Chapters 9 and 10 and the datum tolerance is included in the component position tolerance specification as in Example 6-1. For given size datum features, the allowable tolerance is calculated from formulae 9-1 and 9-2 for internal surfaces and from formulae 9-1 and 9-4 for external surfaces.

Pattern Location Dimensions

Pattern location dimensions are dimensions between datum surfaces and a pattern of features, such as dimensions *A* and *B* of Figures 6-21 a and b. Although the practice is now obsolete, many drawing can still be found with toleranced dimensions between the edges (implied datums) and the lines of true position for the feature pattern.

When non-functional edges are involved, it is reasonable to add a tolerance between a hole pattern and the edge. This allows the pattern of holes to be displaced relative to the part outline and makes manufacturing of the component much easier in some cases.

Specifications such as that of Figure 6-21a mean that the dimensional tolerances apply between the implied datum surfaces and the theoretical lines of true position. Each feature can still deviate within the pattern in accordance with the position tolerance specification while the pattern location tolerances can be though of as permitting movement of either the entire pattern, or of the datum surfaces. Figure 6-22 shows the latter interpretation

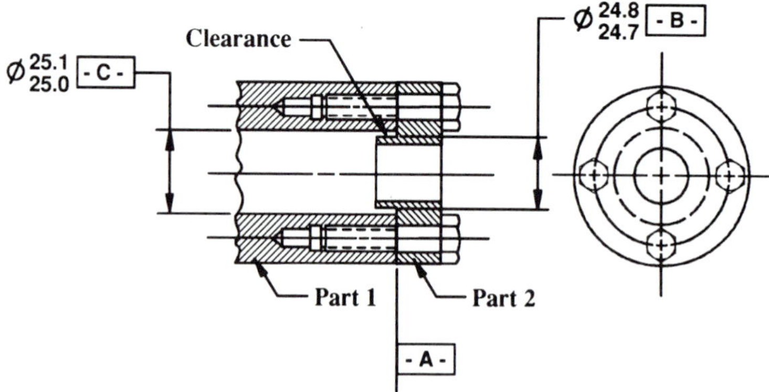

Figure 6-20. Assembly with datum features which assemble with clearance

Example 6-1

Calculate a balanced tolerance for the pilot diameter for the components in Figure 6-20.

Solution Solving by the use of the fixed fastener formula 10-6a:

$$T_B = \frac{H - F}{2} = \frac{25.0 - 24.8}{2} = 0.1$$

The resulting specification is shown below. As the pilot is related to the pattern which has both size and location tolerance, the pattern becomes a virtual size datum for the pilot.

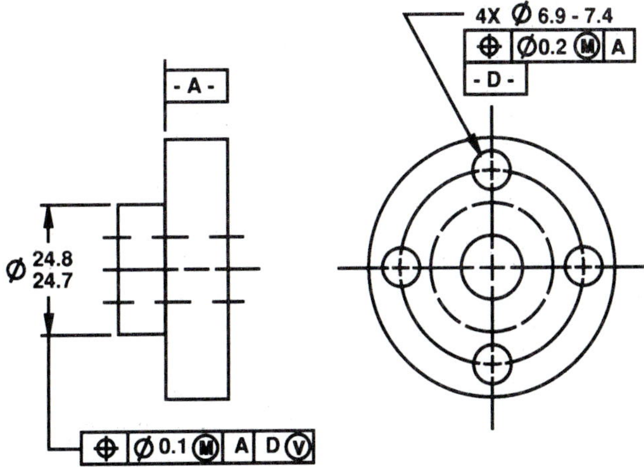

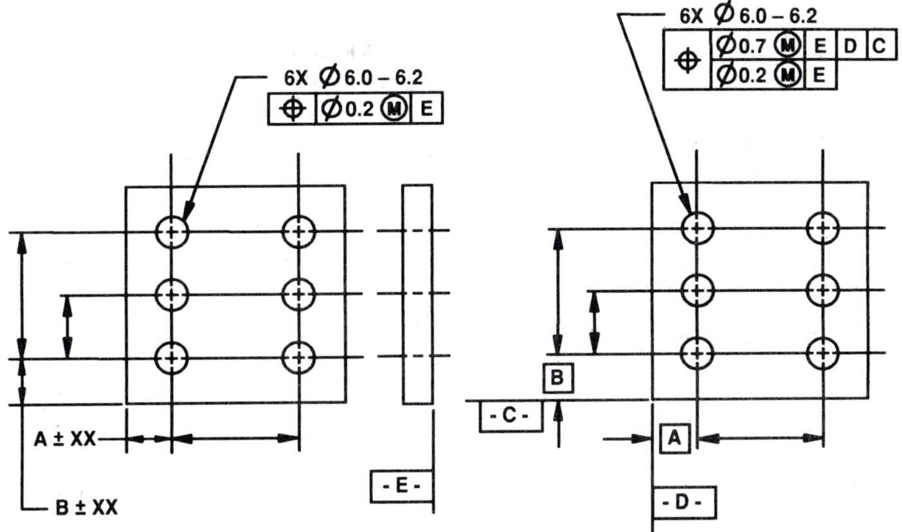

a) Pattern location dimensions

b) Composite position tolerance

Figure 6-21. Pattern location methods

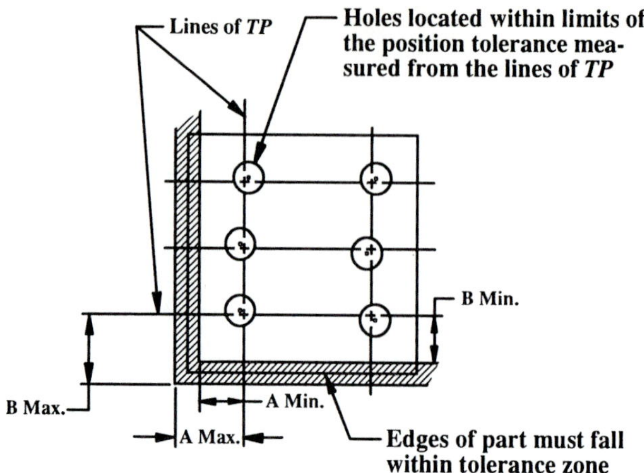

Figure 6-22. Interpretation of pattern location dimensions

for the component of Figure 6-21a. In practice, gages were built to inspect these edges by scribing the tolerance zone band on the gage base. Practically speaking, such gages were difficult to use unless the tolerances were very large.

The current practice is to use either no pattern location tolerance as in Figure 6-3, or to apply two levels of tolerance control to the features, as in Figure 6-21b. The portion of the specification carried over from Figure 6-21a represents tolerance control *within the pattern.* Since the holes now comply with the *within pattern* requirement, the difference between the requirement to the part edge datums and the *within pattern* requirement represents a *pattern location tolerance.* The merit of this approach lies with the ease and accuracy that it can be gaged. The *within pattern* requirement is checked by six pins at true position that are sized 0.2 under the MMC hole size and perpendicular to the base plane *E*. By placing removable datums at D and *C* and using additional pins 0.7 under MMC, the pattern location is easily and accurately checked. The two size pins are frequently combined into step pins, thereby reducing the gage complexity.

DATUM MATRIX

Datum systems range from very simple to extremely complex. Figure 6-23 displays a matrix of the many factors that affect theory, application, documentation, and use of datums. Keeping these categories clear in your mind helps to resolve the confusion that arises on complex parts and in discussions where different parties approach the same topic from a different focus.

SUMMARY

Datums are the origins from which locations are measured and the reference system for Orientation controls. Datums exist in precision equipment.

Positioning a component requires that its 6 degrees of freedom be limited. These are

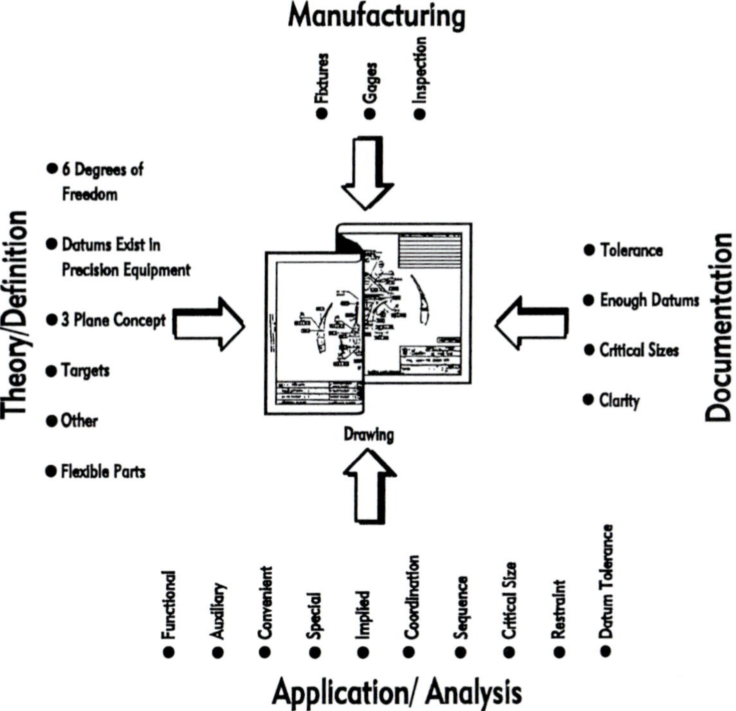

Figure 6-23. Datum Matrix

3 degrees each of translation and rotation. Many methods including the three datum plane method, the target point method, and others, can be used to constrain a component. Flexible parts require restraint limited by applied force or free state dimensional variation.

Datums are either specified by product drawings or implied by functional requirements. Datum types include:

1. Functional datums.
2. Convenient datums, which are features selected to be used as datums because of their convenience.
3. Special datums, which are features added to a component for use as a datum, especially as locators for fixtures and gages.

Datum features of size require consideration of critical size. When functional datums, such as pilots, have a positive clearance at all times, the feature may be assigned a position tolerance.

Composite position tolerance is the currently accepted method of controlling pattern location where tolerance is desired.

Chapter 7

Geometric Tolerances

The broad group of tolerances applied to control part geometry, other than tolerances applied directly to individual dimensions, is called GD&T, or as suggested by this author *geometric datums and tolerances*. The symbols used to designate geometric tolerances were introduced in Chapter 5.

In this chapter, we will cover, in detail, the meaning of each tolerance category. Table 7-1 relists the various geometric tolerance types (see column 3). Because these individual tolerance types have developed in a piecemeal fashion, and because the subject is immensely complex at the detail level, Table 7-1 has regrouped the various tolerances into categories of tolerance which perform similar control and in which certain relationships, such as critical size and datum requirements, are consistent.

As shown in Figure 7-1 every geometric tolerance is either a tolerance band between two parallel lines or between two parallel surfaces, or a three-dimensional zone shaped like its feature (e.g., a cylinder for a round hole). This simple, but important, insight helps to avoid misinterpretations that suggest that other limits can be imposed. This insight also helps to highlight the second important role of datums, that of defining the tolerance class by the way in which datums are used for that class of tolerances.

It should also be recognized that GD&T applies no restrictions upon the surface or line toleranced other than it be contained entirely within the tolerance zone. Hence, a toleranced surface can be wavy, have discontinuities, or any other type variation as long as it is contained within the band of any applied tolerance. As the controlled surface may vary anywhere within its tolerance zone, it is sometimes necessary to apply two or more tolerances to a common surface or feature and a hierarchy is established in Table 7-1 to make sure that such multiple requirements are meaningful. As *profile tolerances* can be used in

Table 7-1. Geometric Tolerances

CLASS	CONTROLS	TOLERANCE TYPE	SYMBOL	CRITICAL SIZE APPLICABLE: LINES/ SURFACES	CRITICAL SIZE APPLICABLE: FEATURES OF SIZE	DATUMS REQUIRED	HIERARCHY
FORM	LINE ELEMENTS	STRAIGHTNESS	⏤	NO	YES	NO	1 (LOW)
		PROFILE OF A LINE	⌒	NO	NO		
		ROUNDNESS	○				
	SURFACES	FLATNESS	⏥	NO	NO	NO	2
		PROFILE OF A SURFACE	⌓				
		CYLINDRICITY	⌭				
ORIENTATION	SURFACES	ATTITUDE -PARALLELISM	∥	NO	YES	USUALLY A BASE REFERENCE PLANE	3
		-PERPENDICULARITY	⊥				
		-ANGULARITY	∠				
	LINE ELEMENTS	RUNOUT -CIRCULAR	↗	NO	NO	AXIS OF ROTATION	
	SURFACES	-TOTAL	⌰				
LOCATION	LINE ELEMENTS	PROFILE OF LINE	⌒	NO	DOES NOT APPLY	BASE REFERENCE PLANE TO COMPLETE RESTRAINT	4 (HIGH)
	SURFACES	PROFILE OF SURFACE	⌓				
	FEATURES OF SIZE	POSITION	⌖	DOES NOT APPLY	YES		
		CONCENTRICITY	◎		NO		
	TOTAL GEOMETRY	TOLERANCED DIMENSIONS		NO	NO	IMPLIED	

different ways, they have been repeated in two classes to permit an appropriate hierarchy to be established.

PERFECT FORM AT MMC

One of the first things to understand when considering geometric tolerances is how they relate to individual size dimensions and tolerances. Although interpretation can be difficult, certain guidelines are available:

1. Limits denoted by size tolerance are considered to be absolute, hence rounding of measurements above or below the stated size limits is not allowed. Any dimensional limit should be considered as if it had an infinite number of zeros following the last given decimal place.

2. The virtual size, or boundary of a component that is constructed by its MMC size limits, is of perfect geometric form and should not be violated. This is an obvious extension of the absoluteness of individual size dimensions extended to three-dimensional form. This provides an insight to the hierarchy ranking of Table 7-1. If geometric variation is limited by size tolerances, then geometric tolerances must have smaller values, or refine the size

Every geometric tolerance is either:

1. A tolerance band between two parallel lines.

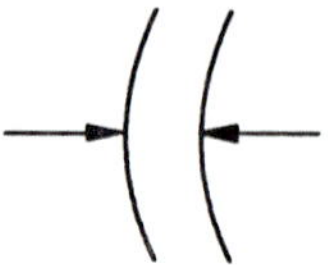

2. A tolerance band between two parallel planes.

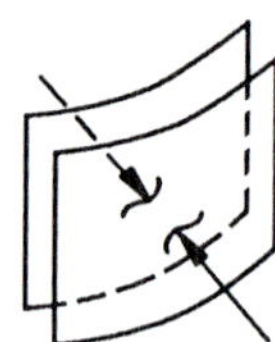

3. A 3-D tolerance zone shaped like the feature.

Figure 7-1. Insight!

limits, to be meaningful. As we shall consider further, this hierarchy or ranking also applies between tolerance classes. Hence, it is possible, for example, to have a *flatness tolerance* contained within a *parallelism tolerance,* contained within a *size tolerance*. Note that to be meaningful tighter tolerances are required as we move lower in the hierarchy.

3. Tolerance standards do not hold that the LMC limit is an opposing boundary of perfect geometric form, although that would seem to be appropriate. Part of the problem is that an MMC boundary is external and manageable from a measurement viewpoint, whereas an LMC boundary is much more difficult to conceptualize and measure. Instead, the standards define the LMC limits as applicable only in a cross section or individual measurement. Although size dimensions control both MMC and LMC boundaries, in practice geometric tolerances are used both to refine these limits and to clarify specific tolerance needs when they are important and boundary control may not be ensured by size limits. Since there is no easy way to inspect a part to determine if the boundary is violated, an appropriate geometric tolerance can provide the focus to ensure that specific needs are fulfilled.

When geometric tolerances are noted to apply at the MMC limit the perfect form at MMC requirement is superseded. This occurs when tolerances generate a boundary type control as with certain applications of *straightness, perpendicularity,* and so on.

FORM TOLERANCE

The first class of tolerances that we will review are *form tolerances.* This group is low in the hierarchy and does not involve complex relations between features. As such, datum relationships are not required and, with one exception, critical size relationships do not exist. Within the class, a division occurs between the tolerances controlling line elements and those controlling surfaces.

Straightness

Straightness requires that a line element be sufficiently straight so as to lie between two parallel lines separated at a distance equal to the tolerance value. The shape of the part is not relevant except that it must have straight line elements. Figure 7-2[1] shows a *straightness tolerance* requirement and its interpretation. Figure 7-3 shows straightness applied to a roller bearing. Here the straightness ensures proper contact with the bearing's mating surface. As we see here, a relationship to the MMC size is implied so as to not violate the perfect form at MMC.

Figure 7-4 shows a straightness tolerance applied to an MMC critical size diameter. Hence, the tolerance applies at the largest diameter and increases if the part is smaller. One way to visualize such applications is to generate a virtual size boundary from the MMC diameter and the Straightness tolerance. Alternately it may be thought of as a cylindrical tolerance zone applied to the axis of the part. As will be seen later, this is analogous to the alternate concepts of axis or surface control for position tolerance.

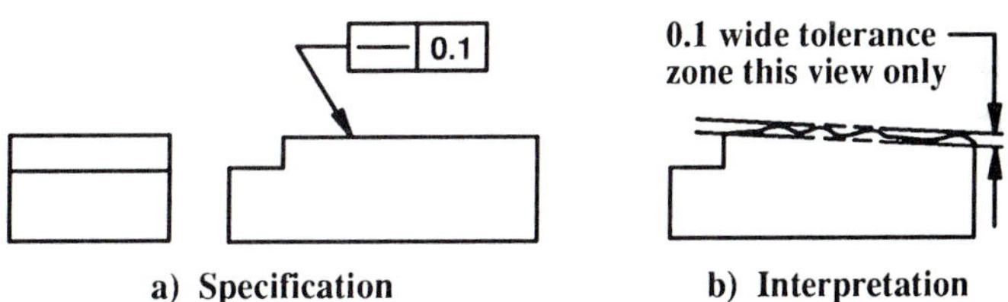

Each longitudinal line element of the surface must lie between two parallel lines 0.1 apart on the right hand view.

Figure 7-2. Straightness

[1] Many of the illustrations in this chapter are reproduced from or based upon the standard "Dimensioning and Tolerancing" ANSI Y14.5M—1982. Figures 7-2, 7-3, 7-4, 7-6, 7-7, 7-9, 7-11, 712, 7-13, and 7-18 are direct reproductions or are based upon figures in the standard and are reproduced with the permission of the American Society of Mechanical Engineers, United Engineering Center, New York, NY.

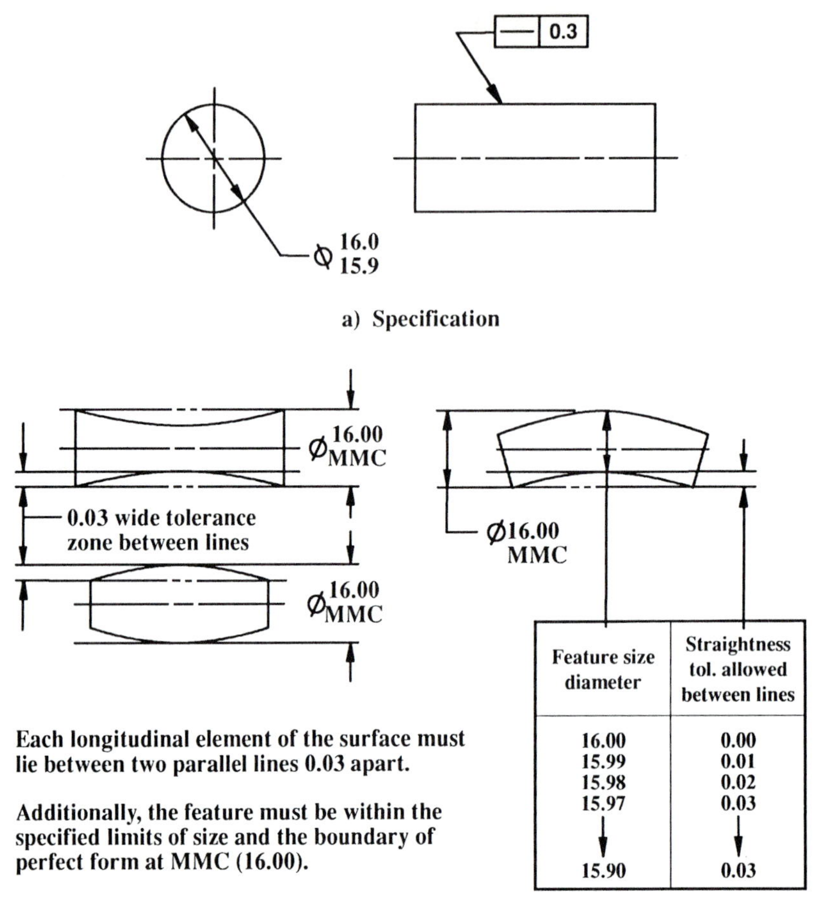

0.3
Ø 16.0
15.9
a) Specification
Ø16.00
MMC
0.03 wide tolerance
zone between lines
Ø16.00
MMC
Ø16.00
MMC
Each longitudinal element of the surface must
lie between two parallel lines 0.03 apart.
Additionally, the feature must be within the
specified limits of size and the boundary of
perfect form at MMC (16.00).
b) Interpretation

Feature size diameter	Straightness tol. allowed between lines
16.00	0.00
15.99	0.01
15.98	0.02
15.97	0.03
↓	↓
15.90	0.03

Figure 7-3. Straightness

Profile of a Line

Profile of a line, when used as a form tolerance, is a generalization of straightness, the difference being the use of curved lines for control as opposed to straight lines. Figure 7-5 illustrates a typical application. This type of tolerance has limited use and is applied when the cross section requirement is more stringent than the surface requirement.

Roundness

Roundness is a further expansion of the concept of element control. It is like profile of a line except the curve is closed to itself. Figure 7-6 applies Roundness control to several parts shapes. Again, note that the control applies to a cross section.

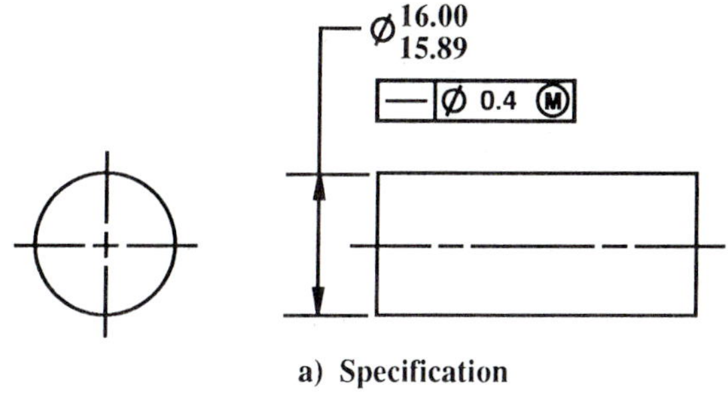

a) Specification

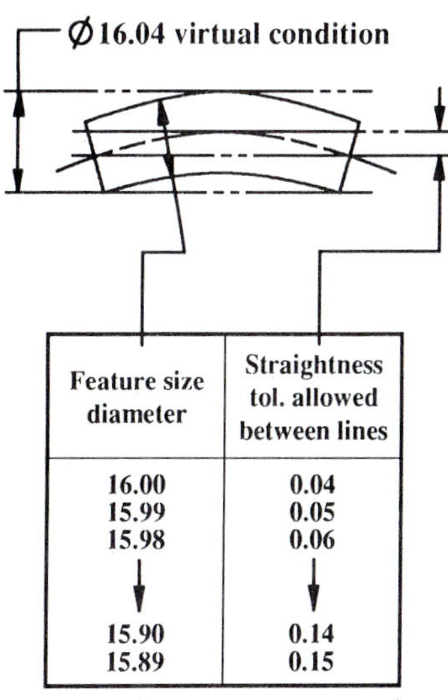

Feature size diameter	Straightness tol. allowed between lines
16.00	0.04
15.99	0.05
15.98	0.06
↓	↓
15.90	0.14
15.89	0.15

This derived axis or center line of the actual feature must lie within a cylindrical tolerance zone of 0.04 diameter at MMC. As the feature departs from MMC, an increase in the straightness tolerance is allowed which is equal to the amount of such departure. Additionally, each circular element of the surface must be within the specified limits of size.

b) Interpretation

Figure 7-4. MMC Straightness

Flatness

Flatness is the first of the surface type of form tolerances. It is the three-dimensional equivalent of straightness. Figure 7-7 illustrates Flatness control applied to the part initially shown in Figure 7-2. The straight line tolerance band of Figure 7-2 has been replaced by two parallel planes separated at a distance equal to the tolerance value stated. Flatness is one of the easier form tolerances to relate to its application. It is often used on surfaces that

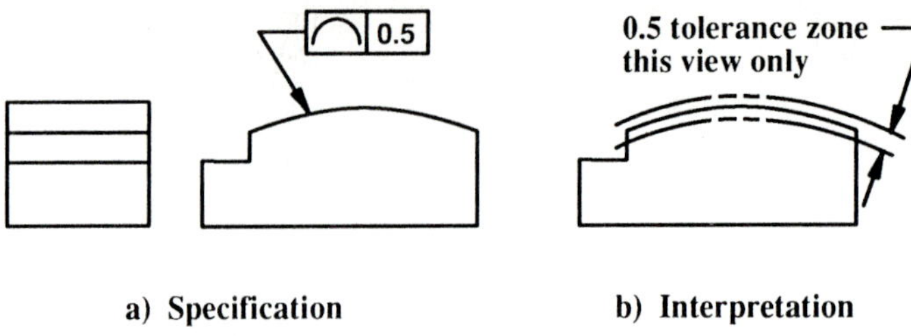
0.5
0.5 tolerance zone
this view only
a) Specification
b) Interpretation
Each line element of the surface must lie between two parallel curved lines 0.1 apart on the right hand view.

Figure 7-5. Profile of a line form tolerance

mount face to face, typically with a gasket to achieve a seal. The top surface of an automobile engine block and the mating head surface are typical examples of applied flatness. Tolerance values here should relate to the amount that the gasket can compress and seal effectively.

Profile of a Surface

Profile of a surface, when used as a form tolerance, is a generalization of the flatness tolerance in the same manner that profile of a line is a generalization of straightness. Figure 7-8 illustrates profile of a surface control applied to the part originally shown in Figure 7-5. The

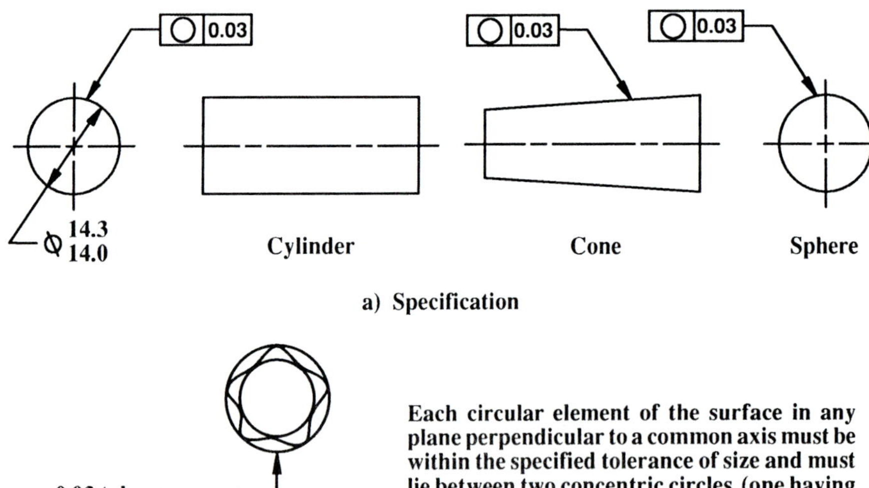
0.03
0.03
0.03
14.3
14.0
Cylinder
Cone
Sphere
a) Specification
0.03 tolerance zone
this view only
Each circular element of the surface in any plane perpendicular to a common axis must be within the specified tolerance of size and must lie between two concentric circles (one having a radius of 0.03 larger than the other).
b) Interpretation

Figure 7-6. Roundness

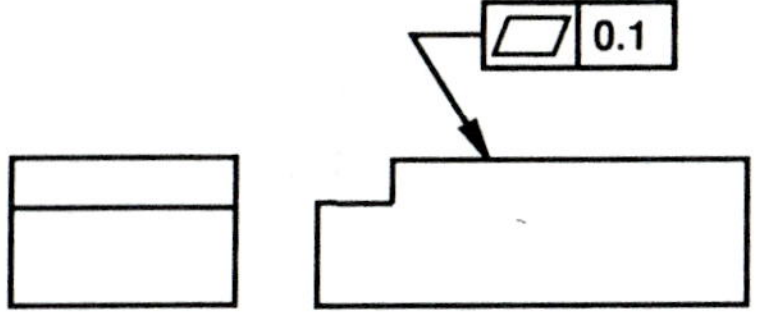

a) Specification

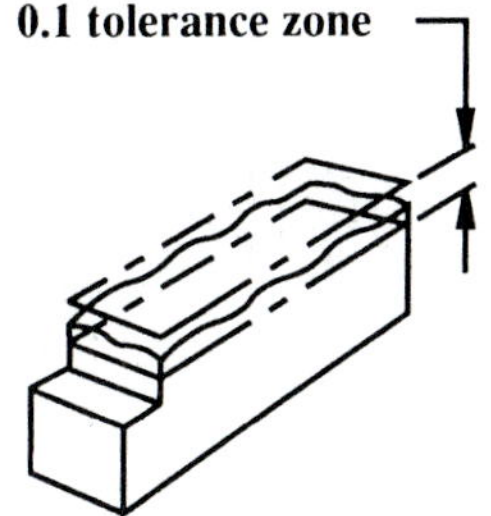

b) Interpretation

The surface must lie between two parallel planes 0.1 apart.

Figure 7-7. Flatness

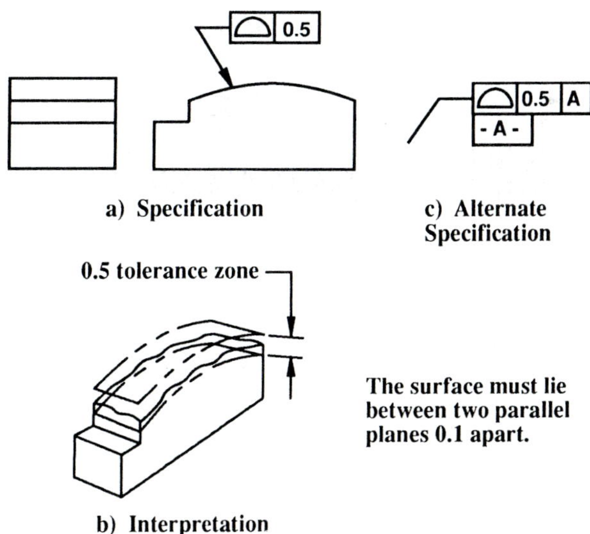

Figure 7-8. Profile of a surface form tolerance

curved line tolerance band of Figure 7-5 has been replaced by two parallel curved planes separated at a distance equal to the stated tolerance value.

You can occasionally find profile tolerances which were intended as form tolerances but which include complete datum specifications. This should always be avoided as it makes the tolerance appear to be a location tolerance. This is one of the serious deficiencies of the current standards in that the various possible uses of profile tolerance can be confusing and careless application of datums can change the intended meaning. One large company's standards require datums in profile form tolerance specifications as shown in Figure 7-8c. The obvious implication is that the tolerance is self datumed. The author does not recommend this practice as it is inconsistent with other form tolerance specification practices, such as flatness. Also complex parts would require the reservation of many datum letters, a seemingly small problem, but one of significance in the real word of preparing product drawings.

Cylindricity

Cylindricity is the three-dimensional equivalent of roundness and the tolerance zone is created by extended the round tolerance band along the axis of a cylinder as shown in Figure 7-9. This type tolerance is seen infrequently, but is appropriate for components such as roller bearings.

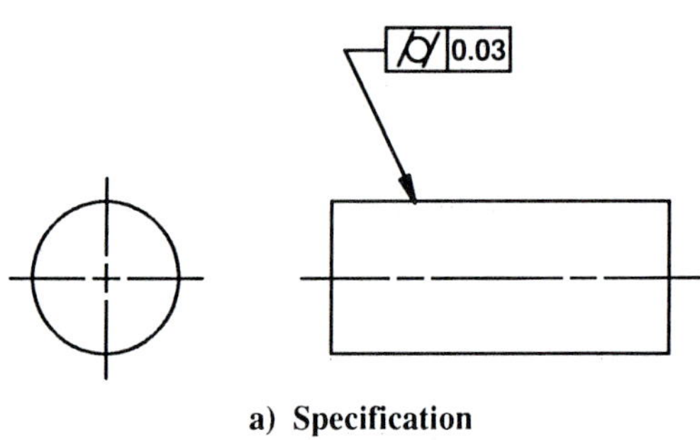

a) Specification

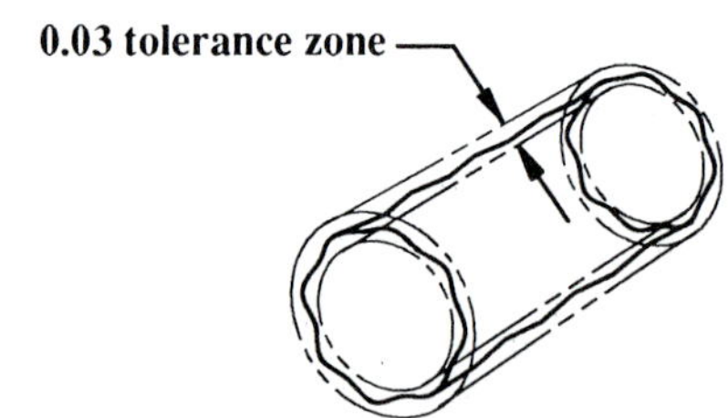

b) Interpretation

The cylindrical surface must be within the specified tolerance of size and must lie between two concentric cylinders, one having a radius 0.03 larger than the other.

Figure 7-9. Cylindricity

ORIENTATION TOLERANCE

Orientation tolerance is the second class of tolerances. These tolerances are similar to surface form tolerances except that they include a datum or datum set. The datums do not locate the tolerance zone, rather they orient it so that it remains parallel to the nominal surface. As orientation tolerances are above the form tolerance class in hierarchy, they also control form unless a tighter form tolerance is applied.

Attitude Tolerance

Attitude tolerance groups together three tolerance types, *parallelism, perpendicularity,* and *angularity.* When applied to surfaces, these tolerances exercise control much like surface form tolerances by requiring the controlled surface to lie between two planes separated by a distance equal to the stated tolerance value. The difference vis-à-vis surface form tolerance is the introduction of a datum which orients, but does not locate, the tolerance zones. Thus, as shown in Figure 7-10b the tolerance zones float with the controlled surface while maintaining their orientation to the datum, in this case parallel to plane A.

The interesting point about this group of tolerances is that they are identical except for the position of their datum. This is illustrated in Figure 7-10 by retaining the same controlled surface and moving the datum. Thus, in Figure 7-10a, the tolerance zone is oriented parallel to plane A, while in Figure 7-10c it is oriented perpendicular to plane B, and in Figure 7-10d it is oriented at a basic angle to plane C.

As shown in Figure 7-11, attitude tolerance zones may also assume a cylindrical shape. This is appropriate for applications to features of size, such as projections, holes, and the like. As shown, the diameter symbol is included in the tolerance specification to clarify the intent. This example also includes a critical size MMC designation. Hence, the tolerance applies at the largest diameter of the pin and increases when the pin is smaller in size.

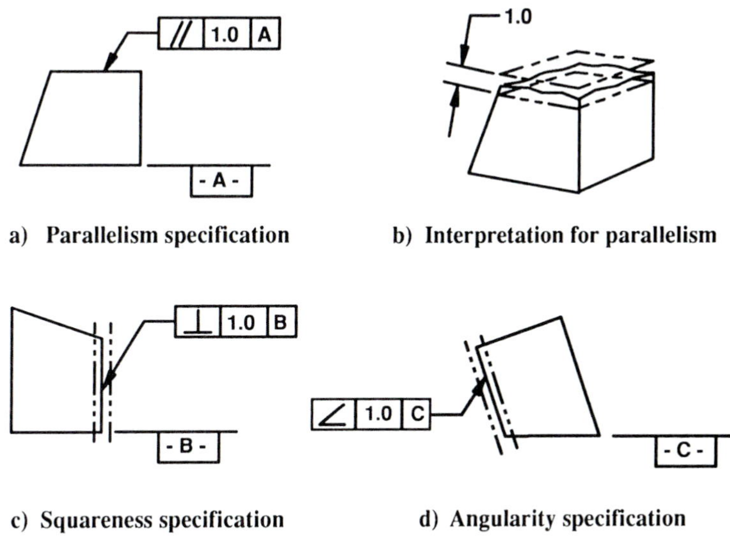

Figure 7-10. Attitude tolerances

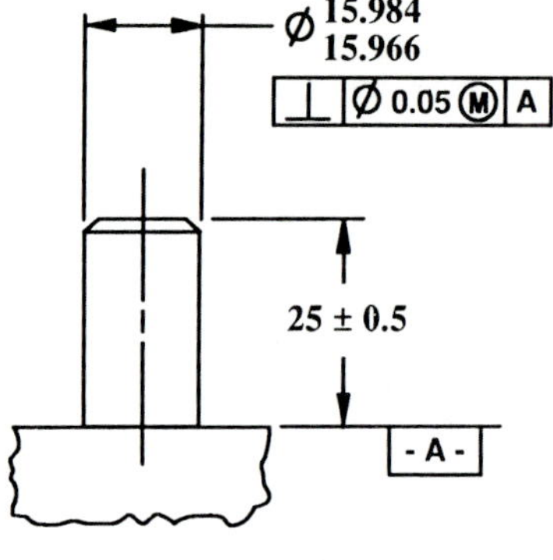

Where the feature is at maximum material condition (15.984), the maximum perpendicularity tolerance is 0.05 diameter. Where the feature departs from its MMC size, an increase in the perpendicularity tolerance is allowed which is equal to the amount of such departure. Additionally, the feature axis must be within the specified tolerance of location.

a) Specification

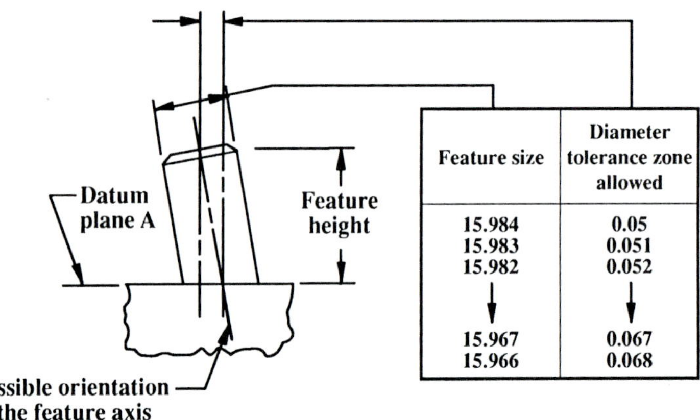

Feature size	Diameter tolerance zone allowed
15.984	0.05
15.983	0.051
15.982	0.052
↓	↓
15.967	0.067
15.966	0.068

b) Interpretation

Figure 7-11.

The combined MMC critical size and the allowed perpendicularity tolerance creates a virtual size of 16.034.

Lastly, Figure 7-10, like the figures used in the various standards, shows attitude tolerance applied only to flat surfaces. This implied restriction is not necessary as the same logic applies to the control of curved surfaces. Hence, such a restriction should not be assumed and this author recommends the angularity symbol as the specification for curved surfaces. The logic used is that a curved surface has a basic angular relationship to its datum system. Figure 7-21 shows attitude tolerance control for a curved surface on a complex part.

Runout

There are two types of *runout* control: *circular* and *total.* Their use is selected as to best maintain design intent. Circular runout is shown in Figure 7-12a. It is a line element control similar to roundness control, but is oriented to a datum axis and, hence, adds eccentricity error to the roundness error. This runout tolerance band applies independently to individual

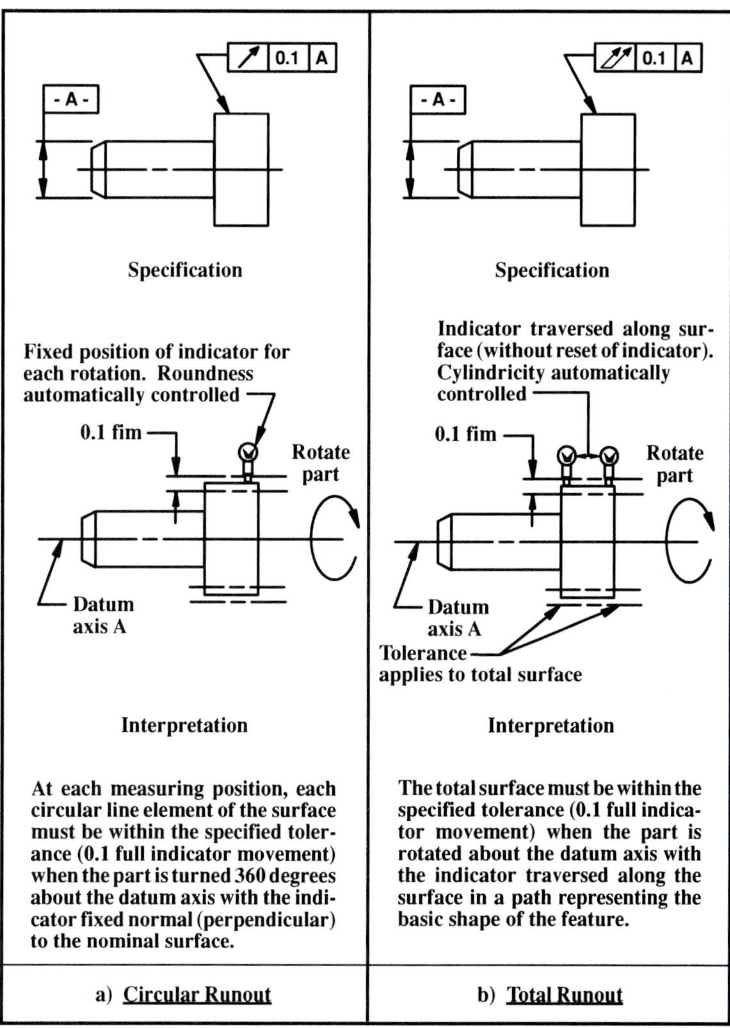

Figure 7-12. Run-out

sections, and may be specified to apply only to a specific location on a surface, while total runout is a surface control similar to cylindricity with a datum axis and its tolerance band applies to the total length of the surface being controlled. Figure 7-12b defines total runout for the part introduced in Figure 7-12a. Occasionally one sees a runout specification containing a critical size modifier, such as MMC. Although such an application was previously acceptable, it does not conform to current standards and a position tolerance should be used.

LOCATION TOLERANCE

This tolerance class tops the hierarchy, sharing this ranking with toleranced dimensions. Consistent with this ranking, *location tolerances* also control orientation and form unless tighter orientation or form tolerances are applied. Complex and complete datum sets are involved. Position tolerance applies to patterns of features. Several chapters will be devoted to this topic. Our focus here will be on profile tolerance used as a location tolerance.

Profile Tolerances as Location Tolerances

Profile tolerances used as location tolerances are the geometric tolerance equivalent of a toleranced dimension. Their original application (see Figure 7-13a) was to apply a uniform tolerance zone to an irregular shape. Such a surface, if based upon many individual dimension and tolerances, will have a non-uniform tolerance width.

As shown in the figure and defined by the ANSI standards, the tolerance zone is centered on the nominal surface as located by the specified datums. If the tolerance zone is unilateral, as in Figures 7-13b and 7-13c, the ANSI standard requires that the unilateral or unbalanced tolerance be designated graphically by phantom part outline. In practice, particularly on drawings of thin parts, this graphical method does not work well as it is difficult to determine the relationship of the phantom line to the part.

Figure 7-14a shows a symbolic method used by the author to convey unbalanced tolerance requirements. Figure 7-14b shows this method applied to the unilaterally toleranced examples of Figure 7-13. In such cases, it would not be necessary to show the phantom part outlines.

Figure 7-15 shows a complex location tolerance application of both profile of a line and profile of a surface tolerance to control a flange which is subject to springback. The profile of a line tolerance reflects the fact that the part is more accurate and subject to bilateral error near the bend, while the profile of a surface tolerance is larger and unbalanced in the direction of the expected springback. Use of this method permits the smaller tolerance to be used in stack-ups when parts are welded flange to flange and then straightened.

Recently, there has been significant growth in the use of GD&T to define allowable variability of components drawn by *computer-aided design* (CAD). In some systems, no dimensions exist and the part is defined by a math data base in computer memory. In many cases, profile tolerances are added to perform the role previously accomplished by toleranced dimensions. Hence, the use of the profile tolerance category is becoming extensive. In such cases, rules previously applied to dimensioning now apply to the GD&T requirements. Figure 7-16 shows a simplified version of a specification that this author has encountered several times. When sketched out in an equivalent dimensional form it is clearly seen to be double dimensioned, a classic unacceptable dimensioning error.

Concentricity Tolerances

Concentricity tolerances will be discussed quite briefly as their current usefulness is minimal. Concentricity is a coaxial control method. It was used extensively in the past for items such as the alignment of various diameters of a turned shaft. In these applications, the specification usually stated a *full indicator reading* (FIR) or *total indicator reading* (TIR) concentricity value. This implies a surface control identical to the current runout tolerance concepts.

As will be seen in Chapter 10, certain fasteners are described well by MMC critical size relations and position tolerances are the current preferred method of control, replacing

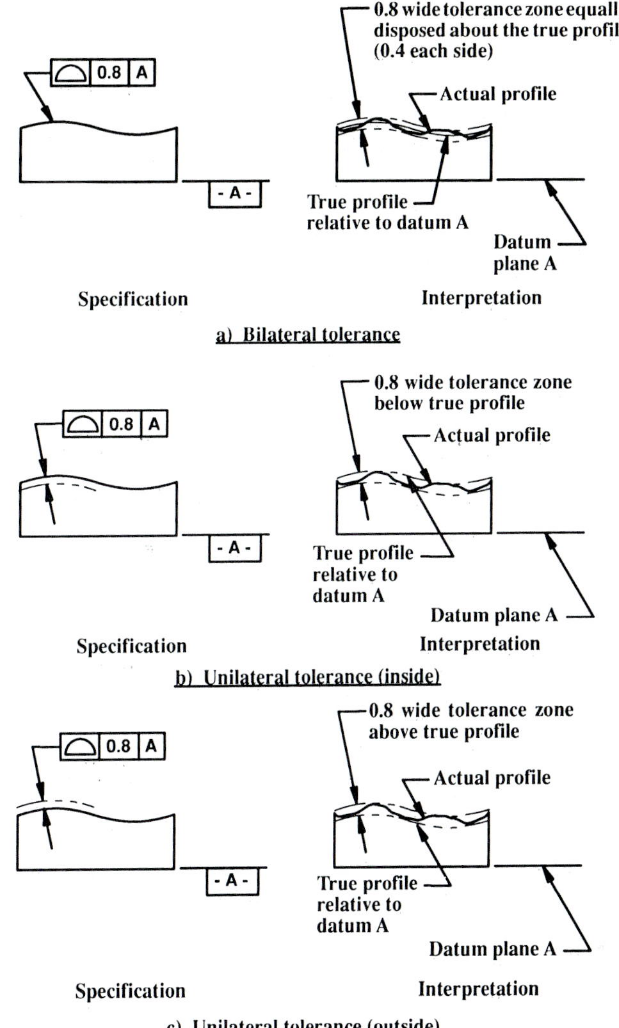

Figure 7-13. Profile of a surface location tolerance

older MMC concentricity requirements. Y14.5M still shows concentricity used for RFS control of the axes of nominally coaxial features. Again, position tolerance with an RFS critical size can be used. In either case, measurement of RFS coaxial features is most difficult.

COMPLEX SPECIFICATIONS

Complex specifications occur in the form of composite and compound tolerances. *A composite tolerance occurs when two tolerances of the same type are applied, and a single symbol is used* as in Figure 7-17. This will occur most frequently with position tolerance, as

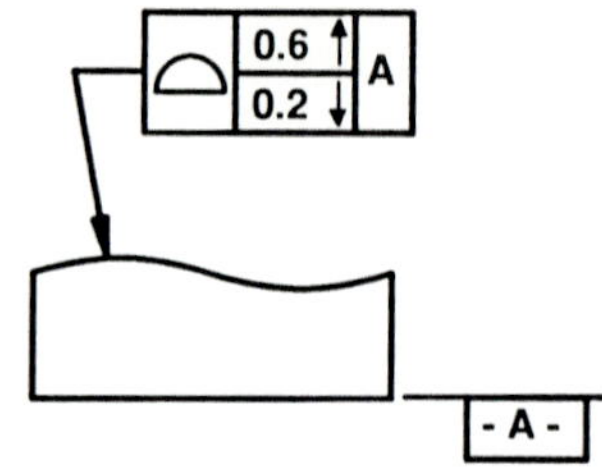

a) Unbalanced Specification

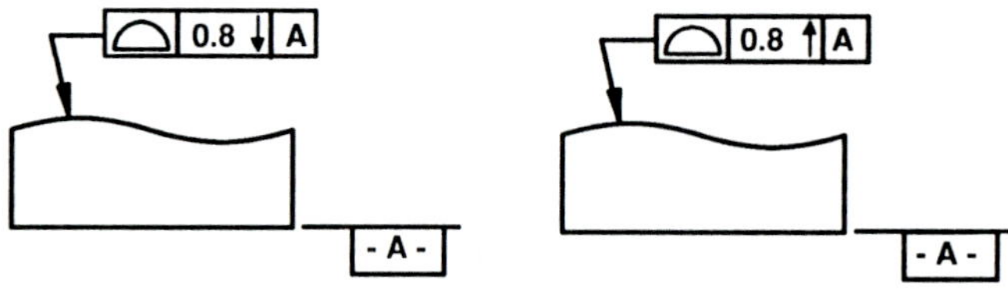

b) Unilateral Specifications

Figure 7-14. Profile tolerance centering

in Figure 7-17, and with profile tolerance that describes both form and location tolerances. The composite position tolerance was discussed in Chapter 6 on datums.

Compound tolerances are shown in Figures 7-18 and 7-19. Figure 7-19 shows a profile location control with reduced runout and further reduced flatness. The important issue here is that the tolerances have reduced values as we progress lower in the hierarchy so as to be meaningful. With the exception of position tolerance applications, compound tolerances will be experienced more frequently than composite tolerances, especially if the simplified approach outlined later in the chapter is accepted.

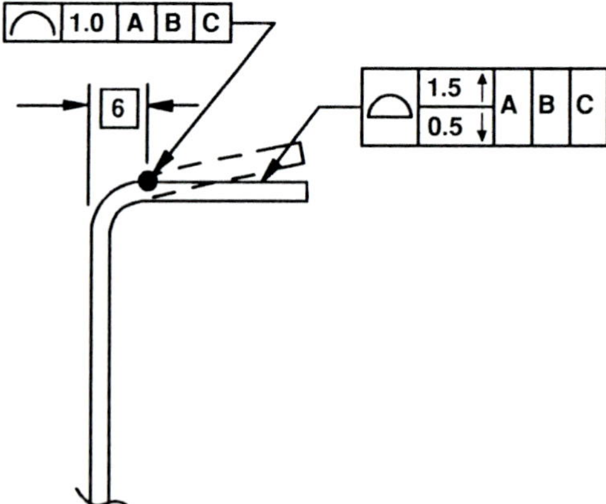

Figure 7-15. Complex application of location tolerance

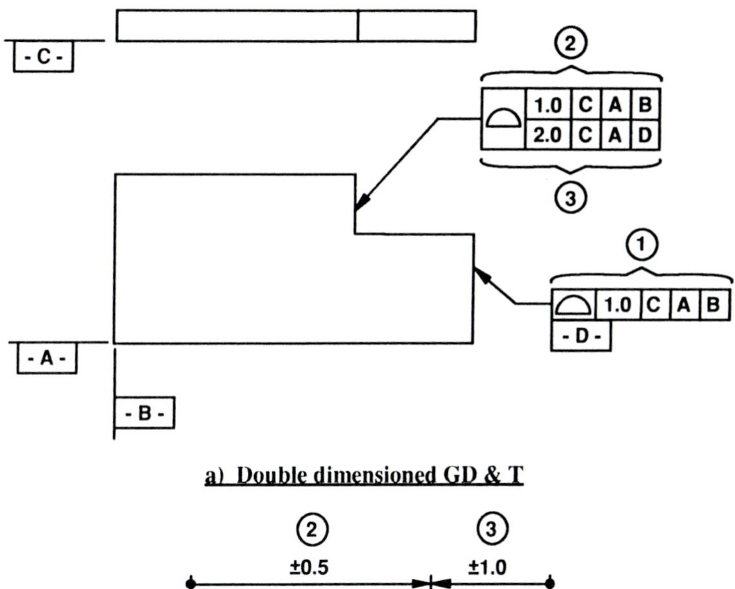

a) Double dimensioned GD & T

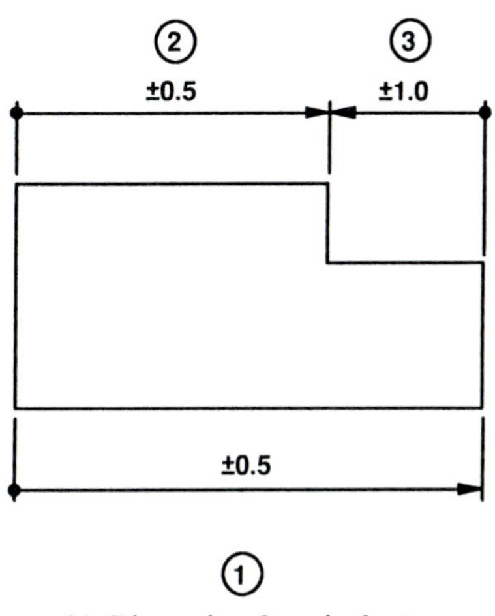

b) Dimensional equivalent

Figure 7-16.

Form tolerance may have a secondary *unit basis* length or area requirement like that shown in Figure 7-20. In all such cases, the unit basis tolerance is interpreted like a standard form tolerance except that the tolerance zone length or area is reduced. The unit requirement applies in any possible position. This may be thought of as floating the tolerance zone (side to side in Figure 7-20). One common misinterpretation is that the tolerance is a ratio that may be applied to a local length or area by multiplying the two values. This is clearly a violation of the definition and the insight gained in Figure 7-1. Unit basis tolerances are both restrictive and difficult to measure. As such, they should be applied only when the product function clearly justifies their use.

Figure 7-21 shows a multiple tolerance requirement for an automobile door. Here the location profile tolerance used would result in an unacceptable door installation if the atti-

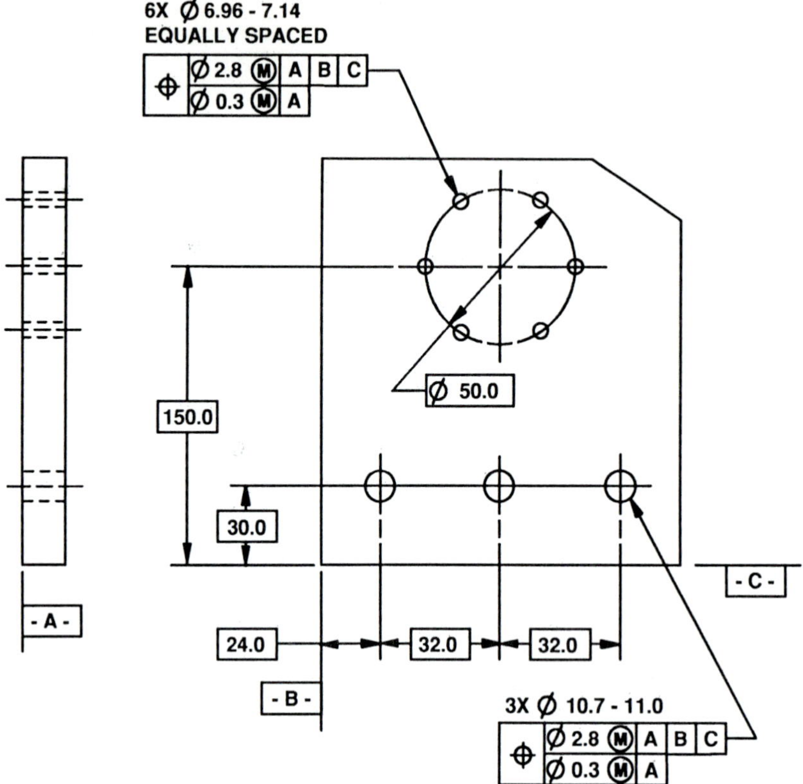

Figure 7-17.

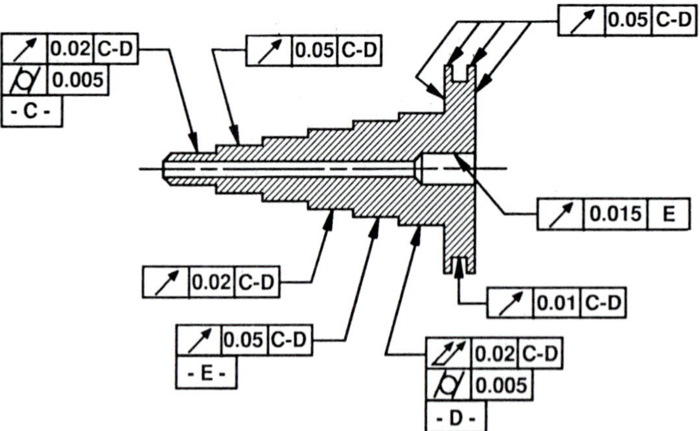

Figure 7-18.

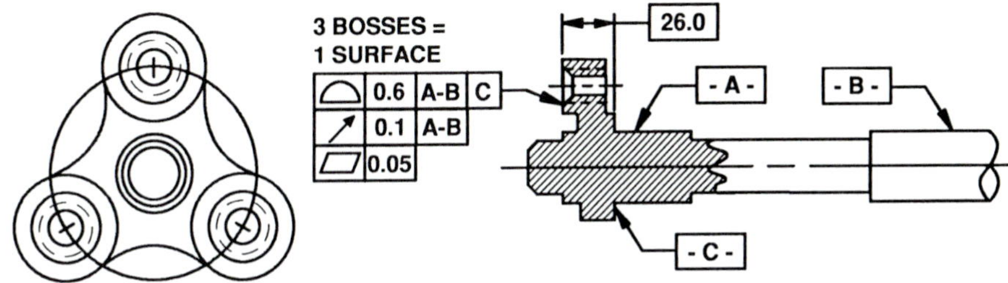

Figure 7-19.

tude of the door edge was such that it fell diagonally within the location tolerance zone, thereby causing the gap to be A or V shaped. The potential error is restricted to a smaller value by the applied attitude (angularity symbol) control. As previously noted, the current standards do not show attitude control of a complex surface, however, it is the author's opinion that this *stretching* of the standards is an important move toward a more universal concept. Other means to define attitude of complex surfaces have been seen. The most troubling of these is yet another profile tolerance. Such uses of profile tolerance to define location, attitude, and form become so confusing that some drawings contain footnotes to explain the intent. Clearly the symbol system must stand alone, hence, this author recommends the angularity symbol as the general attitude control symbol.

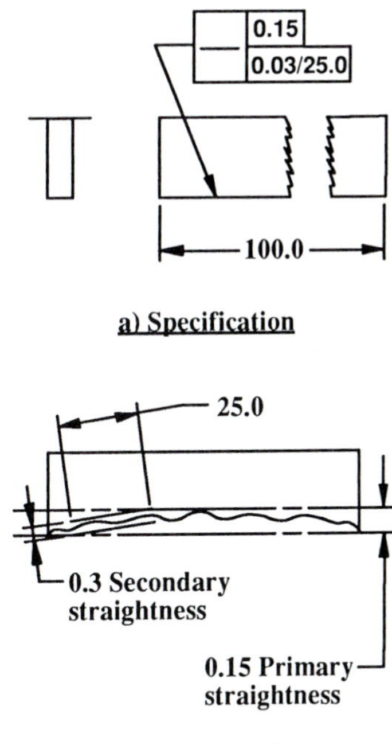

Figure 7-20. Unit basis tolerance

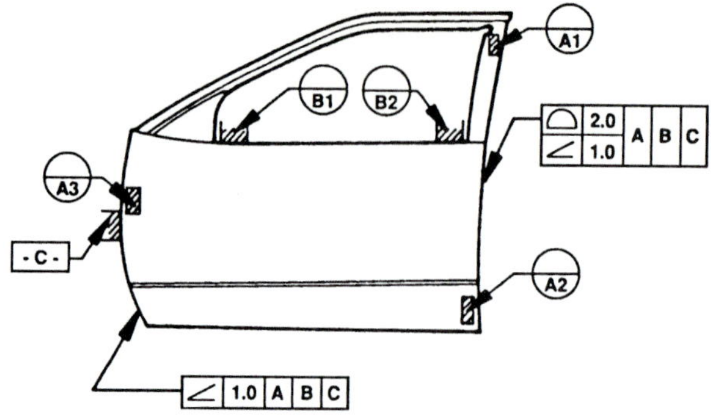

a) Specification

Door edges must lie within 1.0 wide tolerance zones that have the same attitude as the nominal surfaces. These tolerance zones can float from side to side but must be contained in the location tolerance zone, if present.

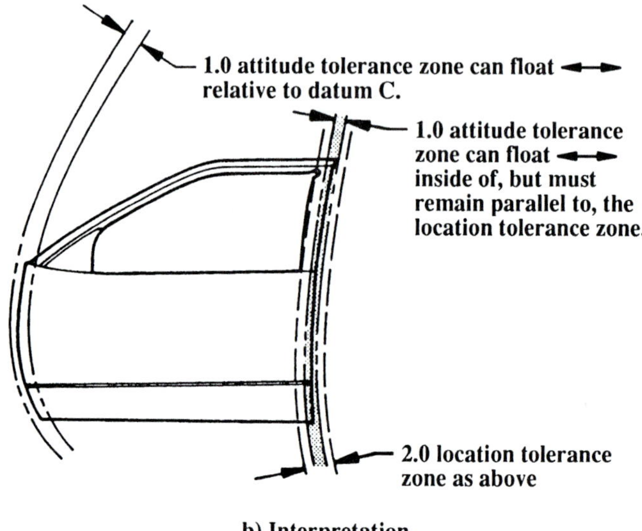

b) Interpretation

Figure 7-21. Attitude tolerance for curved surface

Application

As we will soon see position tolerance values can easily be calculated. Other geometric tolerances do not yield values so readily. As noted earlier, flatness may relate to gasket compressibility. Profile tolerances used as location tolerances may result from, or be used in, a stack-up calculation. Virtual size of MMC critical tolerances can be compared to mating parts. For other tolerances, such as runout and roundness, it may be more difficult to determine functional limits. Many geometric tolerance values are based upon experience or empirical methods. Judgment and review of similar products will usually yield a satisfactory tolerance value. Example 7-1 illustrates how GD&T can be applied to a complex part, in this case an automobile door.

Example 7-1

The assembly requirements (to adjacent panels) are:

Gaps : 5 ± 1.0.
Edges: Parallel within 1.0.
Form within 0.5.

Datums were selected at A, B, and C. Tolerances for surfaces 1 to 3 are selected by considering the need for location, attitude, and form control on each. Note that we start with the highest level control.

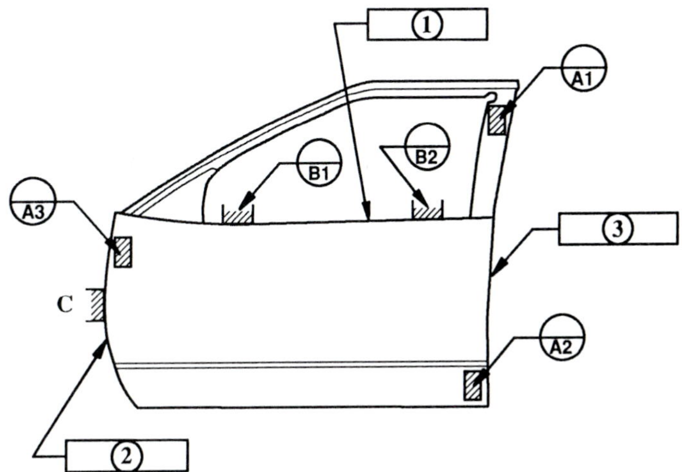

SURFACE 1

Location: Not required—located by the *B* datums.
Attitude: Not required—controlled by the *B* datums.
Form: 0.5 edge form—an assembly requirement.

7-1a

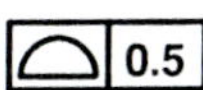

SURFACE 2

Location: Not required—located by the *C* datums.
Attitude: Required to meet the assembly parallelism requirement of 1.0. Assume 1/2 of tolerance applies to door, other 1/2 to adjacent panel.
Form: Not required—Attitude band of 0.5 achieves form control required by assembly requirement.

7-1b

∠	0.5	A	B	C

SURFACE 3

Location: Required to meet assembly gap tolerance of 2.0 (±1). Assume 1/2 of tolerance applied to door, other 1/2 to adjacent panel.

Attitude: Required—same as Surface 2.

7-1c

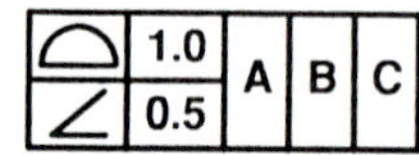

Form: Not required—Same as Surface 2.

Notice that the author simplifies compound tolerance specifications by recording a common datum set only once, similar to the way a single symbol is used in a composite tolerance requirement.

GEOMETRIC TOLERANCE SIMPLIFICATION

As noted in the beginning of this chapter, the various geometric tolerance types developed independently over a long period of time. When we regrouped the tolerances in Table 7-1 to aid our understanding of critical size, datum relationship, and hierarchy, we systemized the geometric tolerance groups. Table 7-2 is a modification of Table 7-1 which has been

Table 7-2. Simplified Geometric Tolerances

CLASS	CONTROLS	TOLERANCE TYPE	SYMBOL	CRITICAL SIZE APPLICABLE		DATUMS REQUIRED	HIERARCHY
				LINES/ SURFACES	FEATURES OF SIZE		
FORM	LINE ELEMENTS	ELEMENT	⌒	NO	YES	OCCASION-ALLY A SECONDARY DATUM	1 (LOW)
FORM	SURFACES	SURFACE	▱	NO	NO	OCCASION-ALLY SECONDARY AND TERTIARY DATUMS	2
ORIENTATION	SURFACES	ATTITUDE	∠	NO	YES	USUALLY A BASE REFERENCE PLANE	3
ORIENTATION	LINE ELEMENTS SURFACES	RUNOUT -CIRCULAR -TOTAL	↗ ⇗	NO	NO	AXIS OF ROTATION	3
LOCATION	LINE ELEMENTS SURFACES	PROFILE	⌓	NO	DOES NOT APPLY	BASE REFERENCE PLANE TO COMPLETE RESTRAINT	4 (HIGH)
LOCATION	FEATURES OF SIZE	POSITION	⌖	DOES NOT APPLY	YES	BASE REFERENCE PLANE TO COMPLETE RESTRAINT	4 (HIGH)
LOCATION	TOTAL GEOMETRY	TOLERANCED DIMENSIONS		NO	NO	IMPLIED	4 (HIGH)

altered in the boxed areas. The changes, which are outlined below, are the authors's recommendations[1] to the standards organizations and major corporations as to how to simplify and improve these tolerance systems:

1. Element tolerance has replaced straightness, profile of a line, and roundness. It provides general control of a line element albeit straight, curved, or closed curve (round). One symbol replaces three. The old profile of a line symbol is suggested as it conveys the general shape. The deletion of profile tolerance in the low hierarchy section removes confusion as to its use in higher levels.

2. Surface tolerance has replaced flatness, profile of a surface, and cylindricity. It is the three-dimensional equivalent of element tolerance and applies to flat, curved, and closed curve (cylindrical) surfaces. One symbol again replace three. A curved analog of the old flatness symbol is suggested as it suggests a general surface. Again, deletion of the profile tolerance from the low level grouping allows it to remain in the higher levels in a more pure form.

3. Attitude tolerance replaces parallelism, perpendicularity, and angularity and the general (angularity) symbol is used. As noted in the discussion of these tolerances, they are identical except for the placement of their datums.

4. Profile tolerance is retained as a location tolerance for general use, except for features of size. The distinction of lines and surfaces is deleted. This author feels that the growth of the use of profile tolerance to support CAD drawings makes it extremely important to clean up the confusion associated with the use of profile tolerance as both a form and a location tolerance.

5. Concentricity is deleted. runout tolerance has replaced the most common of the prior concentricity applications and position tolerance is now preferred for MMC critical size applications. In essence, concentricity has been obsoleted, but not yet withdrawn from the standards.

Thus, we have cleaned up the hierarchy and have defined clear relationships for datum and critical size requirements. The number of tolerance symbols has been reduced from 13 to 7 and, most importantly, no tolerance type appears in more than one class. The elimination of profile tolerance from the form tolerance category also avoids any problem resulting from a datum specification that includes excessive datums. Currently, the only simple distinction between a profile tolerance used as a form tolerance and one used as a location tolerance is the absence of datums on the form tolerance. Ultimately, some datums will be needed for the form tolerance category to achieve accurate coordination of CAD math data section locations for complex surfaces.

Tables 7-3 repeat the figures of affected tolerance types previously shown and contrast current symbology with the proposed simplified method. When no change is required, the figure is not repeated.

[1]The author first suggested that the GD&T symbol system should be simplified in "Element, Surface, and Orient Tolerances," *Design News*, Feb. 16, 1970.

Table 7-3.

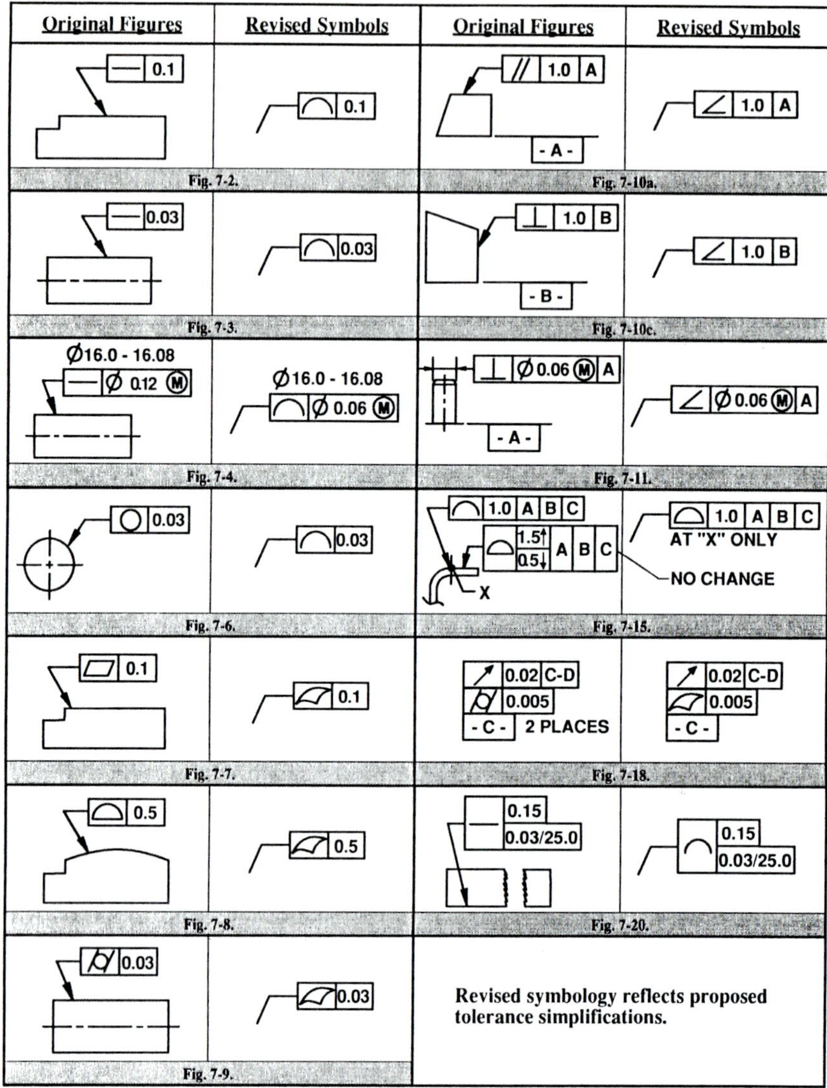

SUMMARY

Geometric tolerances may be grouped as *form, orientation,* and *location tolerances*. These categories help to establish appropriate critical size and datum requirements and tolerance hierarchy

All parts are considered to have perfect form at MMC, however correxponding perfect form is not assumed at LMC. Tolerances noted to apply at a critical size create an exception to this principle.

Form tolerances can be grouped by those that control line elements: *straightness, profile of a line*, and *roundness;* and by those that control surfaces: *flatness, profile of a surface,* and *cylindricity*. These tolerances are represented by a band between two parallel lines or a zone between two parallel planes. No datums are required and critical sizes are seldom used.

Orientation tolerances introduce a datum to orient, but not locate, the surface type tolerance zone. The *attitude tolerances: parallelism, perpendicularity,* and *angularity* are identical except that the datum is moved. *Runout* uses a datum axis and is similar to roundness and circularity.

Location tolerances are high level tolerances and use complex datums and critical size relationships. Profile tolerances assume the role of toleranced dimensions, especially for CAD designed products.

Complex specifications include composite and compound tolerance specifications. form tolerances may also use secondary *unit basis* requirements.

Tolerance values can be difficult to develop and one may need to rely upon judgment, experience, and data from similar parts. Some tolerances, particularly position and profile, can be more rationally calculated.

The author feels that further simplification of the tolerance categories, based upon the grouping of Table 7-2, is necessary for the GD&T system to reach its full maturity.

Chapter 8

The Boundary Concept

Position tolerances[1] may be viewed in two ways: (1) as a boundary limiting the movement of a surface, or (2) as a tolerance zone limiting the movement of the axis of a feature. Both concepts are useful and can be shown to be equivalent; however, the boundary concept is the more flexible of the two systems.

Figure 8-1 illustrates the functional character of the boundary concept. In this application representing a conventional bolted joint, the boundary divides the clearance between the bolt and the hole. Note that the boundary is constructed normal to the interface of the mating parts. Either of the two features can be displaced and the joint will still assemble as long as their surfaces have not moved beyond the boundary.

Because the boundary deals with surfaces, it is obvious that it will always be three-dimensional in nature. It is also easy to see where the boundary should be positioned vertically along the axis of the features by analyzing its relation to the features' function, in this case, the maintenance of clearance to permit assembly. Such considerations can be more difficult, or, at best, less obvious when position tolerances are looked upon as tolerance zones controlling the axes of features. Further, it can be seen that the boundary size is independent of the feature size, except in the case of RFS applications. This provides an important insight into the relation between feature size and position tolerances.

Another important insight is that position tolerance theory and application can be

[1]This chapter and the balance of the book, although partially rewritten, is taken from my original book *Fundamentals of Position Tolerance* and is reprinted with the permission of the original publisher, the Society of Manufacturing Engineers, Dearborn, MI.

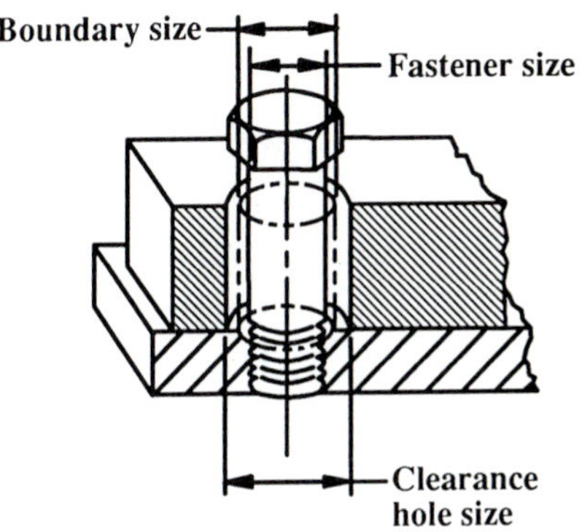

Figure 8-1. Boundary representing limit of movement of hole or bolt to permit assembly

studied by looking at what happens at any one feature. This results from the fact that in patterns of multiple features, each feature must meet the requirements for the pattern to function. For example, in a pattern of clearance holes, each hole must clear its mating fastener for the part to assemble. Hence, studying the clearance at one hole yields the tolerance information for all similar holes. Occasionally, you'll see studies which try to depict the interaction of several features. Such approaches will be complex and incomprehensible and should be discarded in favor of an analysis at a single feature.

DEFINITION OF POSITION TOLERANCE

Based on the boundary concept, *position tolerance* can be defined as two times the distance between the functional surface of a feature and its boundary. The factor of two is introduced to correlate to axis tolerance zones which are total measurements, such as a diameter, while the boundary measurement is along a radius. When the boundary is independent of feature size, the Position tolerance must apply at one feature size, called the *critical size*. When the critical size of a feature is increased or decreased in order to allow for displacements permitted by its position tolerance, the resulting effective size is called *virtual size*. Virtual size is identical to the boundary size for all features except those known as *mounting holes*. Mounting holes will be discussed in Chapter 9.

The boundary can be constructed by drawing its outline parallel to the critically sized feature. It may be either inside or outside of the feature surface, depending upon whether the feature has internal or external surfaces and upon the function of the feature. When a boundary is drawn for a feature, and an abrupt change occurs, such as at a corner, the boundary is completed by extrapolating the contour drawn parallel to the feature until an intersection is attained. As shown in Figure 8-2, the position tolerance at a particular point on the feature surface is measured along a radial line through the point and the center of curvature of the feature contour at that point. This illustration also points out the versatility

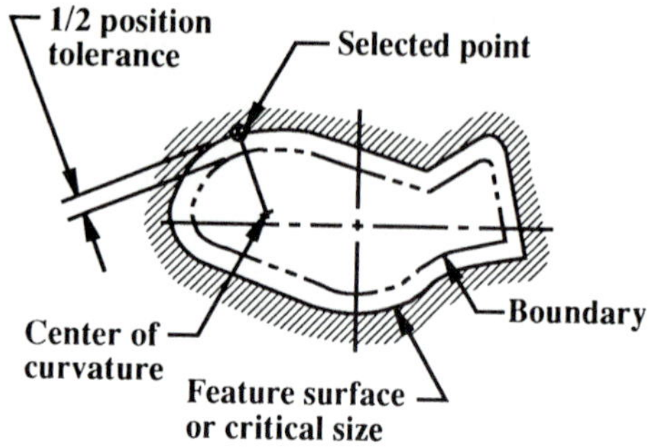

Figure 8-2. Position tolerance is 2 X distance between boundary and critical size

of the boundary concept; that is, its ability to describe by a single concept, and hence a single specification, any type or shape of feature.

CONVERSION TO AXIAL TOLERANCE ZONES

A tolerance zone applied to the axis of a feature can limit the feature position in the same manner that the boundary limits the movement of the feature surface. Because the axis concept of position tolerance is the basis of many inspection techniques, and is a convenient device for expressing position tolerance requirements, its use is always permissible. Since tolerance zone shapes depend upon the feature shape and are difficult to describe by a universal specification, the principle interpretation implied by a position tolerance is best considered to be a boundary concept. When it becomes desirable to use axial tolerance zones, they can be derived from their equivalent boundaries to yield the same control of the feature surface.

Consider, for example, any cross section within the clearance hole of Figure 8-1. If, as in Figure 8-3, the hole is moved in contact with its boundary, the center of the hole traces out the cross section of the equivalent axial tolerance zone. Note for round holes (an certain other regularly shaped features) the relationship between virtual size which is equal to the boundary size in this case) and position tolerance may be written:

$$V_H = H - T_C \qquad (8\text{-}1)$$

where: V_H = Virtual hole size
H = MMC hole size
T_C = Critical position tolerance

Since the cross section shown in Figure 8-3 applies anywhere within the depth of the hole, the tolerance zone becomes three dimensional like the boundary as shown on the left side of Figure 8-4. The other two holes illustrated show that both the tolerance zone and the

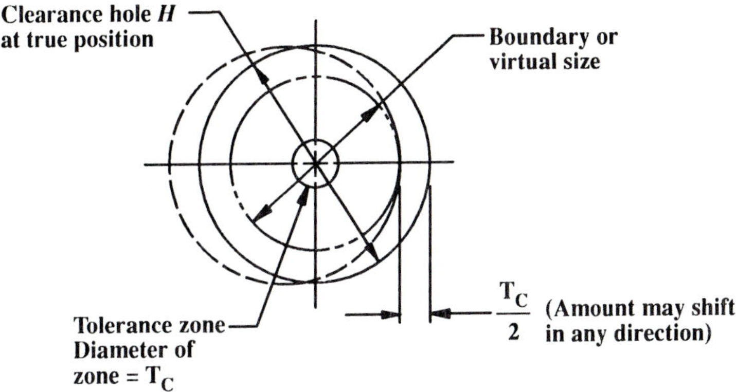

Figure 8-3. Cross section through clearance hole. Moving hole in contact with boundary causes center to generate tolerance zone

boundary exercise the same control and limit both positional variation and perpendicularity of the hole. Figures 8-5 and 8-6 provide different views of the bolt show in Figure 8-1 and illustrate the resulting axial tolerance zone which is derived in the same manner as for the clearance hole. Again, it may seen that the axial tolerance zone and the boundary are compatible and exercise the same control over the bolt. Note that for round fasteners (and certain other regularly shaped features) the relation between virtual size (which is equal to the boundary size in this case) and position tolerance may be written:

$$V_F = F + T_C \tag{8-2}$$

where: F = MMC fastener size

V_F = Virtual fastener size[2]

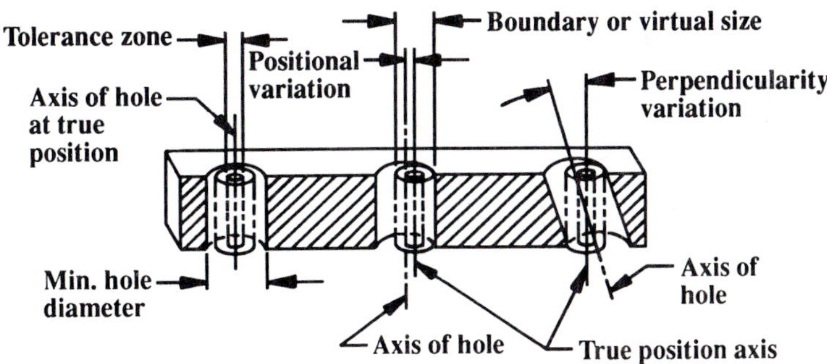

Figure 8-4. Equivalent controls of clearance holes by boundary and axial tolerance zone systems

[2]For assemblies, such as those shown in Figure 8-1, virtual hole size, V_H, and virtual fastener size, V_F, have the same value unless a minimum clearance is maintained. Subscripts are used principally to distinguish between formulae for holes and fasteners.

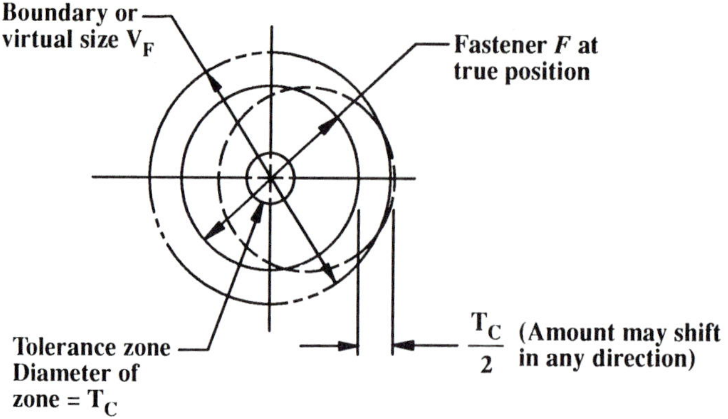

Figure 8-5. Cross section through bolt. Moving bolt in contact with boundary causes center to generate tolerance zone

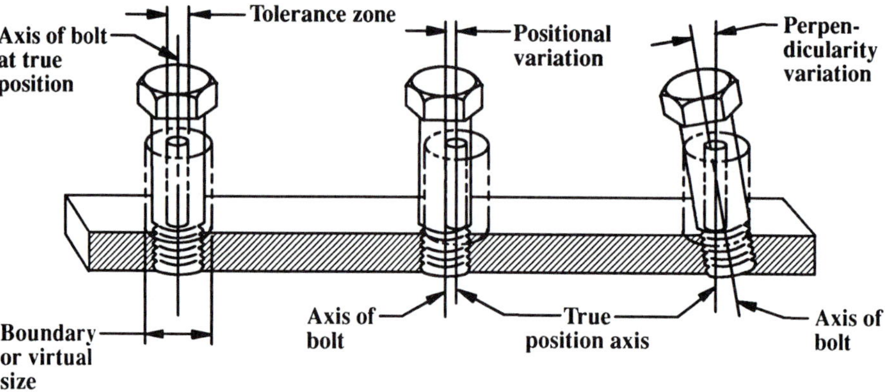

Figure 8-6. Equivalent controls of projecting bolt by boundary and axial tolerance zones systems

SUMMARY

The concept of a boundary limiting the movement of a surface provides the best basis for a position tolerance system because the interaction of surfaces and a boundary is easily visualized and related to functional requirements. Its flexibility results in two important characteristics:

1. The system can be applied equally well to any type or shape of feature.
2. A universal specification format can be adopted.

Position tolerance is defined by the distance between the surface of a critically sized feature and its boundary. Equivalent axial tolerance zones can be substituted for the surface-boundary relation when desired.

Chapter 9

Fundamentals of Position Tolerance

Because of their ability to describe the requirements of interchangeable components, as shown by the boundary concept, position tolerancing techniques are extremely valuable. The comprehensive nature of the system will be seen in this chapter as its fundamental concepts are studied in greater detail. It will become evident that the primary application of position tolerance is to fasteners, and patterns of holes, because no other method so adequately describes the functional requirements of mating hole patterns.

Two types of component features were observed in Figure 8-1 and subsequent figures. The first was the clearance hole, while the second was the fastener and its associated mounting hole. These two types of features will reoccur throughout the remainder of this and later chapters and will provide the basis for much of the material that follows. With either, our goal will be to understand the relation between feature size and location. Once this relation is established, position tolerance values can be determined. In engineering practice, these tolerance values will be specified on the component drawings to control the feature positions, thereby ensuring assembly and interchangeability.

As development of the system proceeds, it will be seen that position tolerances apply individually to each feature within any group, and for this reason, it will be necessary to consider the effects of variation in size and location at only one of the features. This practice significantly reduces the complexity of analyzing any position tolerance application.

PROJECTED HEIGHT FOR MOUNTING HOLES

Both the boundary and the axial tolerance zone for a fastener held by a mounting hole were shown in Figure 8-6. If we were to remove the fastener and consider only the mounting hole, we find that both the boundary and the tolerance zone project above the surface containing the hole as in Figure 9-1. The height of either is designated *projected height* , which is related to the height of the mating part or parts.

Application of Projected Height

At this point, we must examine in greater depth the application of projected height. For a fastener which is inserted into a clearance hole, the height of the projected tolerance zone or boundary is simply the maximum height of the clearance hole as in Figure 9-1. This arrangement is the most common application of projected tolerance.

If the fastener is a stud or an embossment that extends beyond the clearance hole, the assembly requirements differ. Here the mating part is required to assemble over the fasteners, and a projected height equal to the clearance hole height is no longer adequate. This inadequacy results from an improper control of parallelism, as may be seen in Figure 9-2. In this illustration, the mating part cannot be removed although the projected height adequately controls the holes so that assembly or disassembly could be accomplished by installation or removal of the studs through the clearance holes.

The most common approach to this problem is to consider the projected height as being equal to the height of the fastener above the interface surface. This is a simple way to ensure assembly, but is not a true functional description of the fastener requirements. Figure 9-3 indicates the flaw in this approach and is the basis for describing the functional requirements for this type of part.

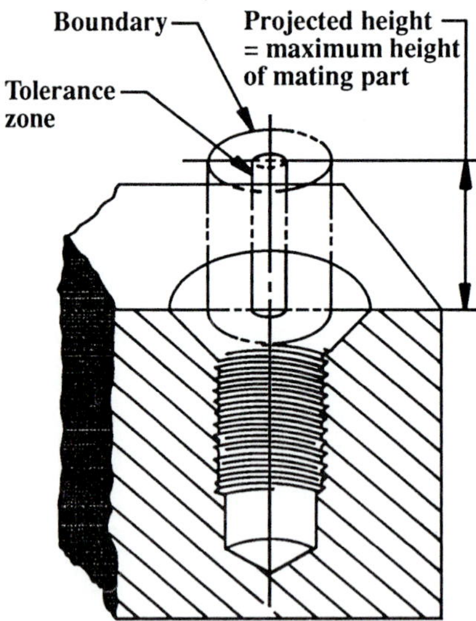

Figure 9-1. Boundary and tolerance zone projected above mounting hole

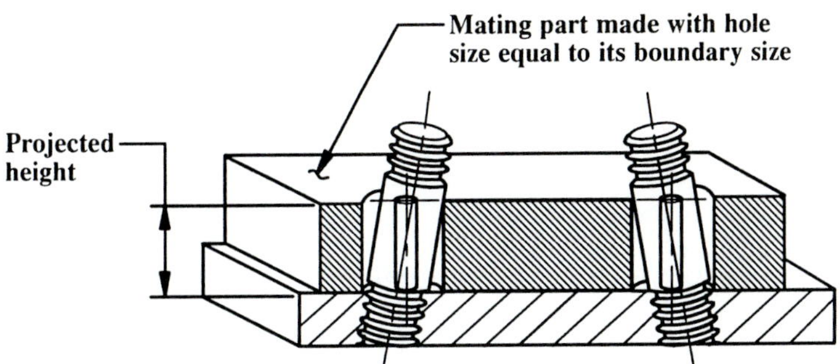

Figure 9-2. Insufficient control of parallelism

Fastener Height[1]

Although the fasteners in Figure 9-3 overhang the holes as in Figure 9-2, the mating part is free to assemble or disassemble. Since the mating part is thin relative to the height of the studs, increasing the height of the projected tolerance imposes a stringent perpendicularity control upon the part when in reality parallelism control is needed. The functional requirements of the parts are satisfied when the body of the fastener falls within the projected boundary or the centerline of the fastener mounting hole falls within the projected tolerance zone, when the upper part (hence the projected height) is placed anywhere within the fastener height. Side-to-side movement of the upper part (or the tolerance zone or the boundaries as a pattern) as it moves over the length of the fastener height is acceptable.

Proper specification of position tolerance for such components will include the value of projected height, or more appropriately, both projected and fastener heights. Note that fastener height specifications must have an associated value of projected height.

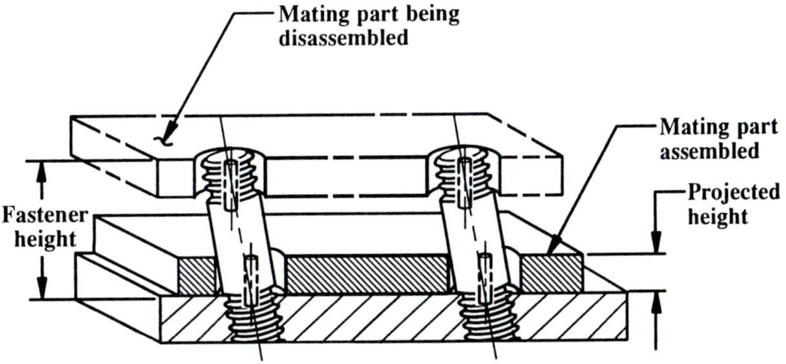

Figure 9-3. Effect of considering fastener height separate from projected height

[1]Fastener height is a recommendation of the author and is not yet included in any published standard.

Multiple Component Assemblies

If more than one part is assembled with a fixed fastener, the projected height is equal to the sum of the maximum heights of the parts when the projected height is used alone. For the projected height-fastener height designation, the projected height should equal the maximum height of the thickest part.

Improper Application of Position Tolerance to Mounting Holes

Although the projected height concept is widely accepted, placement of the tolerance zone within a mounting hole has been recently abandoned with Y14.5M. Therefore, it is important that we consider this concept of tolerance placement and be fully aware of its flaws and the problems associated with it. The principle of analyzing fastener function will enable us to see its fault.

Figure 9-4, which is similar to an illustration used in a major position tolerance standard for many years, shows the tolerance zone for a mounting hole placed coincident with the depth of the hole. The associated text implied that perpendicularity is not adequately controlled by the position tolerance zone. If fastener mounting holes are toleranced in this manner, the extrapolation of an out-of-square hole can cause interference between the fastener and the clearance hole as in Figure 9-5. Note that an imposed perpendicularity tolerance zone cannot stop extrapolation beyond the edge of the position tolerance zone if the perpendicularity tolerance zone is independent of, and free to move within, the position tolerance zone as is required by the hierarchy established in Chapter 7.

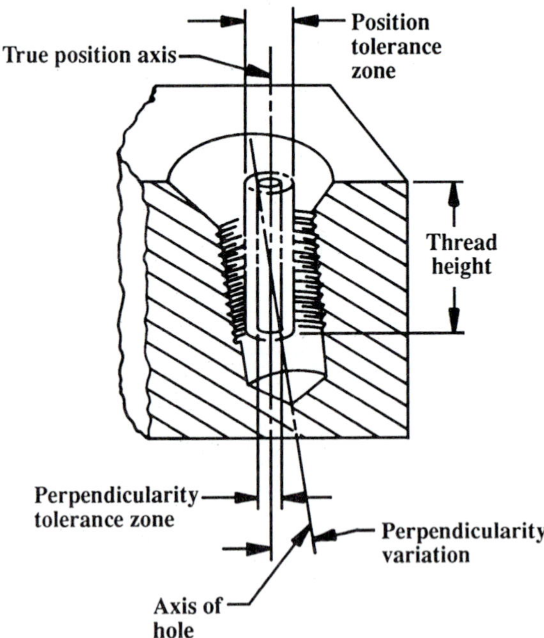

Figure 9-4. Improper positioning of tolerance zones for mounting holes

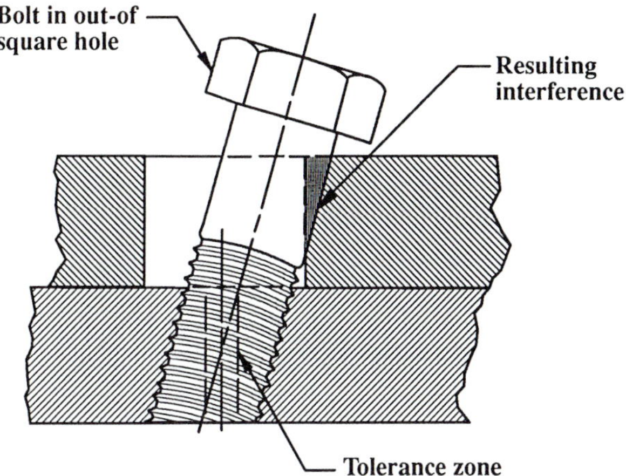

Figure 9-5. Extrapolation of out-of-square bolt. Improperly toleranced mounting hole results in interference

Note that if, as in Figure 9-5, you were to rotate the tipped bolt about its true position axis, it would generate a truncated cone above the interface surface. We have already shown, however, that functional assembly limitations will be described by a cylindrical boundary of tolerance zone (for a round fastener) in this area. Examining this premise in another light, we see that if the clearance hole size is increased to eliminate the interference indicated, the clearance can be absorbed only by a perpendicularity error. This contradicts the principle illustrated by Figure 8-6, that either position or perpendicularity variations may assume the full value of the functionally developed tolerance. Hence, we must conclude that this approach violates the functional requirements of the fastener assembly and is inconsistent with the *total error* concept usually associated with position tolerances.

CRITICAL SIZE

Now that we have considered the geometry of the boundary and the axial tolerance zone, we will determine the effect of feature size variation upon the position tolerance. For any application, we shall determine position tolerances which are acceptable at a critical feature size and then use a specification which allows the position tolerance to increase when the feature is not at its critical size.

Maximum Material Condition (MMC)

The most common type of position tolerance is based on *maximum material condition,* or MMC, size. The intent of this type of Position tolerance is to maintain clearance for a fastener, or other component. As stated in Chapter 8, the boundary size remains constant; hence, additional position tolerance above the specified critical tolerance is allowed when the features are other than MMC size as it cannot adversely affect the desired clearance. As shown in Figure 8-1, the boundary for an internal MMC feature is inside the feature, while it is outside an external MMC feature.

Least Material Condition (LMC)

Some position tolerances are based upon *least material condition,* or LMC, size. The intent of this type of position tolerance is to maintain location of a fastener, or other component. If the features are other than LMC size, additional tolerance is allowed since it will not adversely affect the desired location. As will be shown in Chapter 10, the boundary for LMC holes (the only common LMC application) is outside the holes and has a constant size.

Regardless of Feature Size (RFS)

In applications where a fastener, or other component, and a mounting hole are always aligned, any hole size is critical. Such components are said to be critical *regardless of feature size*, abbreviated RFS. Mounting holes for press fit dowels and for self-tapping screws are examples of this application. Tapped holes are usually MMC applications, but they approach and may be considered to be RFS applications. The boundary for all mounting holes (including both RFS and tapped holes) is above the surface containing the holes because the functional surface is that of the fastener, or other component, held by the mounting hole.

Relation between Size and Location for MMC Critical Holes

As previously noted, this application is associated with fasteners which require clearance to assemble. Assume that an acceptable position tolerance (T_C) is specified for the critical MMC hole size and that the hole is displaced from its true position by one half of that amount (see Figure 9-6). This brings the edge of the hole in contact with the boundary, which represents the space reserved for the mating fastener.

As the hole size increases (approaches LMC), the clearance created will allow the hole to move by an amount equal to one-half the change in hole size (see Equation 9-2) or one-half the hole size tolerance as a limit. When the actual hole size (H_A) is known (as in Figure 9-6), the specified position tolerance (T_C) can be replaced by an allowable tolerance (T_A). The allowable tolerance (see Equation 9-1) includes the increase in tolerance (T_{ADD}) which results from the more favorable hole size. Although round holes are shown in the illustrations, these equations will apply to other geometric shapes as well.

$$T_A = T_C + T_{ADD} \tag{9-1}$$

$$T_{ADD} = H_A - H \tag{9-2}$$

where:
T_A = Allowable position tolerance
T_{ADD} = Additional position tolerance
T_C = Critical position tolerance
H_A = Actual hole size
H = MMC hole size

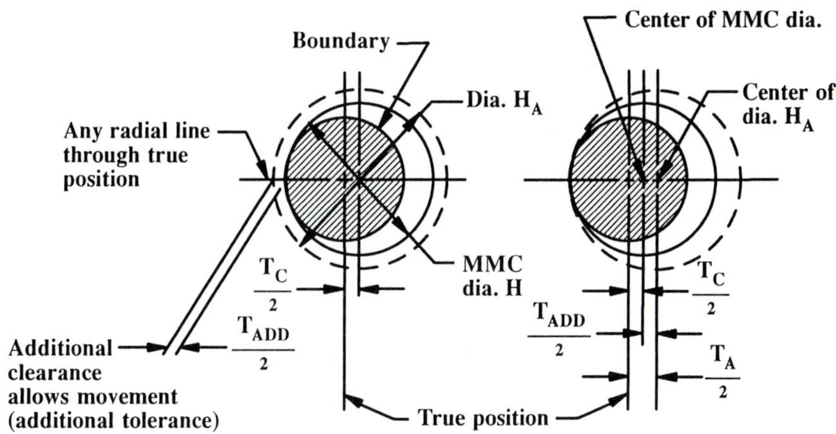

Figure 9-6. Relation between size and location for MMC critical holes

Equations 9-1 and 9-2[2] can be used to generate a chart such as Table 9-1 which yields values for T_A.

Example 9-1

Given a MMC specification of ∅ 0.5 position tolerance, actual hole size = 11.3, MMC hole size = 11.0, find the additional and allowable tolerance.

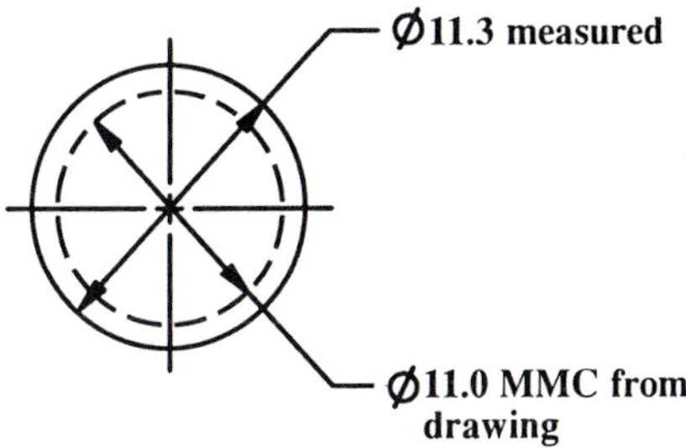

Solution

From Equation 9-2:

$$T_{ADD} = 11.3 - 11.0 = 0.3$$

From Equation 9-1:

$$T_A = 0.5 + 0.3 = 0.8$$

[2]This equation assumes that the holes are geometrically true. Imperfectly shaped holes may have several different values for T_{ADD}.

Table 9-1. Alowable tolerance values (T_A)

Critical position tolerance (T_C)					
0.5	0.6	0.7	0.8	0.9	1.0
0.4	0.5	0.6	0.7	0.8	0.9
0.3	0.4	0.5	0.6	0.7	0.8
0.2	0.3	0.4	0.5	0.6	0.7
0.1	0.2	0.3	0.4	0.5	0.6
0.0	0.1	0.2	0.3	0.4	0.5
	0.1	0.2	0.3	0.4	0.5

Deviation from critical size

Example 9-2

A hole size of 10.8 – 11.3 is specified with a MMC position tolerance of ∅ 0.2. The manufacturing process requires a position tolerance of ∅ 0.4. Calculate the hole size limits that can be used with this position tolerance.

Solution

$$T_{\text{ADD}} = 0.4 - 0.2 = 0.2$$

From Equation 9-2:

$$H_A = T_{\text{ADD}} + H$$

$$H_A + 0.2 + 10.8 = 11.0 \text{ (New MMC diameter)}$$

Resulting hole size limits = ∅ 11.0 – 11.3

One of the most important, but less apparent, advantage of position tolerances is that the manufacturing engineer is allowed to adjust the nominal hole size (within its tolerance range) so as to gain additional position tolerance, when advantageous.

Relation between Size and Location for LMC Critical Holes

This application is associated with fasteners which require positive location; for example, piloted weld nuts. Assume that an acceptable position tolerance (T_C) is specified for the critical LMC hole size, and that the hole is displaced from its true position by one-half of that amount (see Figure 9-7a). This brings the edge of the hole in contact with the boundary, which represents the space within which the fastener must be positioned to assemble with mating parts. As this hole size decreases (approaches MMC), the material added will allow the hole to move by an amount equal to one-half the change in hole size (see Equation 9-3) or one-half the size tolerance as a limit. When the actual hole size (H_A) is known, as in Figure 9-7b, the specified position tolerance (T_C) can be replaced by an allowable tolerance (T_A). The allowable tolerance (see Equation 9-1) includes the increase in tolerance (T_{ADD}) which results from the more favorable hole size. Although round holes are shown in the illustrations, these equations will apply to other geometric shapes as well.

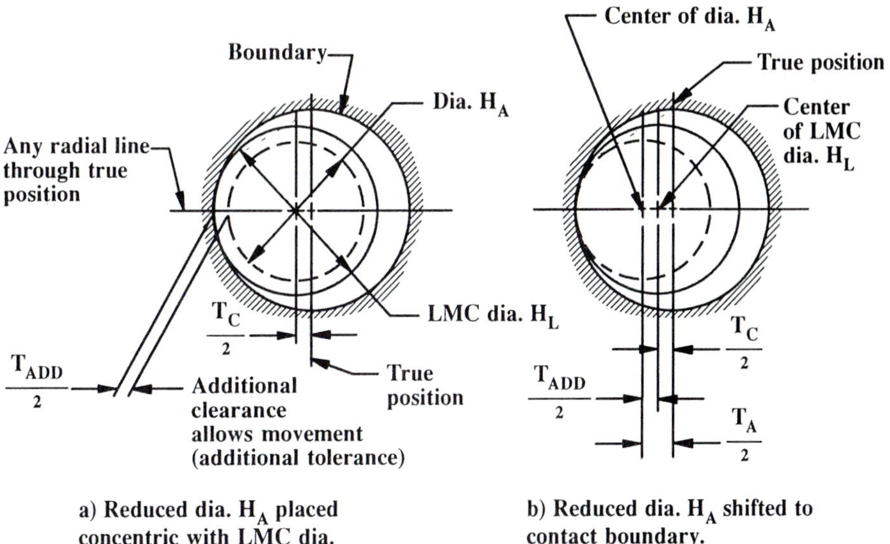

Figure 9-7. Relation between size and location for LMC critical holes

$$T_{ADD} = H_L - H_A \tag{9-3}$$

where: H_L = LMC hole size

T_A for LMC critical hole sizes can also be determined from Table 9-1.

LMC specifications can be adjusted in the same manner as MMC specifications to gain advantageous position tolerance for manufacturing requirements.

Relation between Size and Location for MMC Critical Projections

MMC projections, sometimes called *bosses*, may be used to assemble components by crimping, staking, or upsetting the projection so that it permanently engages its clearance hole. Such a practice is common in the electrical appliance industry.

Projections are functionally similar to fasteners secured by mounting holes, and their tolerance structure is similar to that shown in Figure 8-6. Unless otherwise specified, the boundary or tolerance zone of any projection is assumed to be equal in height to the height of the projection as shown in Figure 9-8. Where a part is assembled over a projection which is relatively long compared to the thickness of the part, the part thickness may be specified as a projected height to make the position tolerance requirements less stringent. In this case, the overall projection height becomes an implied fastener height.

As with clearance holes, projection assembly requirements are based upon the need for sufficient clearance and therefore are associated with an MMC critical size. Since the boundary surrounds a projection, the position tolerance will increase in magnitude as the projection decreases in size. This boundary feature relationship is identical to that developed for LMC critical holes, as can be seen by comparing Figure 9-7, illustrating the LMC

Example 9-3

Given a LMC specification where the actual hole size = 10.3, LMC hole size = 10.7, and the specified position tolerance = ∅ 0.3. Find the additional and allowable position tolerance.

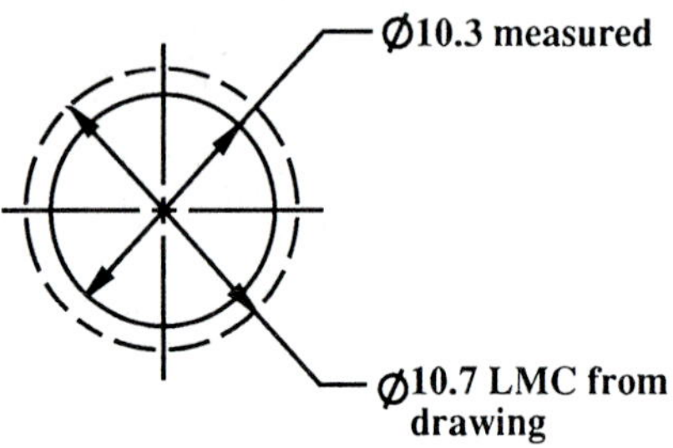

Solution

From Equation 9-3:

$$T_{ADD} = 10.7 - 10.3 = 0.4$$

From Equation 9-1:

$$T_A = 0.3 + 0.4 = 0.7$$

Example 9-4

A hole size of 11.0 – 11.3 is specified with a LMC position tolerance of ∅ 0.15. The manufacturing process requires a position tolerance of ∅ 0.3. Calculate the hole size limits that can be used with this position tolerance.

Solution

$$T_{ADD} = 0.3 - 0.15 = 0.15$$

From Equation 9-3:

$$H_A = H_L - T_{ADD}$$
$$H_A = 11.3 - 0.15 = 11.15 \text{ (New LMC diameter)}$$

Resulting hole size limits = ∅ 11.0 – 11.15

hole application, with Figure 9-9, which illustrates the MMC projection application. In both cases, the boundary is placed outside the feature. Equation 9-1 defines the relation between allowable, critical, and additional tolerance for MMC projections, as well as for the previously developed applications. Table 9-1 may also be used to obtain this value for projections. Equation 9-4 defines the increase in tolerance (T_{ADD}). Although round projections are shown in this illustration, these equations also apply to other geometric shapes. As with other applications, projection size may be adjusted to gain advantageous position tolerance for manufacturing requirements.

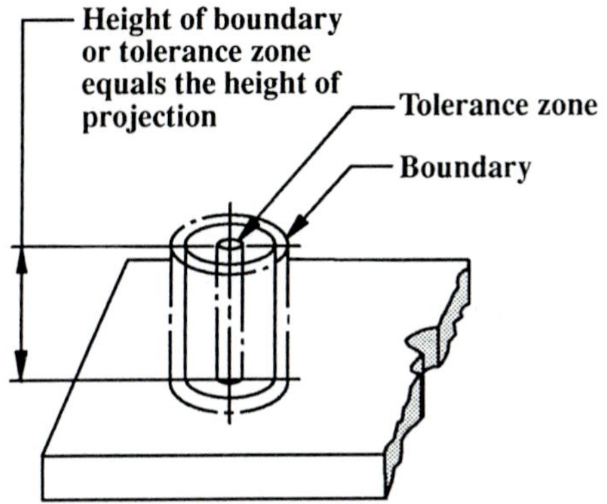

Figure 9-8. Tolerance geometry for MMC critical projections

$$T_{ADD} = F - F_A \tag{9-4}$$

where: F = MMC projection size
F_A = Actual projection size

Conversion of Critical Sizes

Theoretically, it is possible to convert a specification based upon one critical size to another. The conversion of RFS or LMC critical sizes to an MMC basis is sometimes suggested so as to take advantage of the easier gaging methods of the MMC feature. It has been

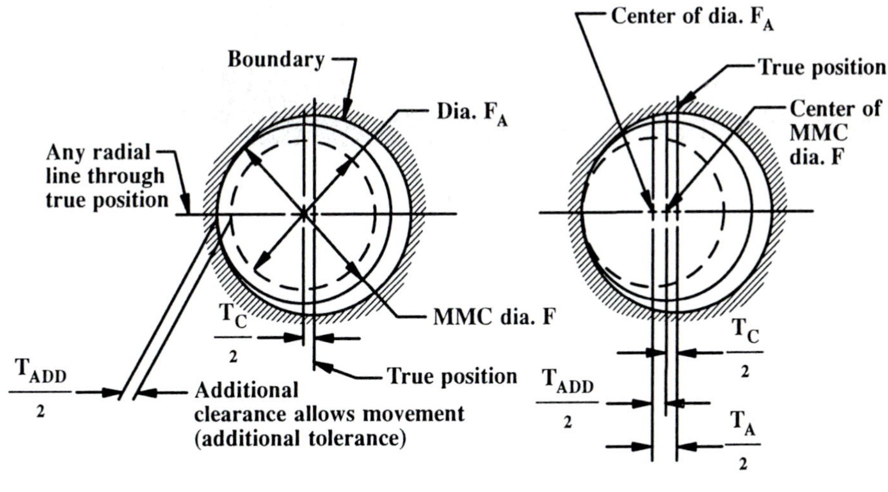

Figure 9-9. Relation between size and location for MMC critical projections

suggested that this can be accomplished by making the critical size tolerance of the RFS or LMC feature the maximum tolerance of an MMC feature of the same size range. Then, by calculating a negative additional tolerance from the feature size range, this maximum tolerance can be adjusted to obtain a compatible MMC critical size tolerance. This approach suffers from theoretical considerations in one case, and practical considerations in the other. With the LMC application, the principle is impractical as the calculation will result in negative tolerances for the MMC specification in most applications. The RFS application is not sound as it usually deals with projected tolerances.

As may be seen in Chapter 12, the virtual size and, hence, the equivalent MMC tolerance of a feature using projected height specifications is nonuniform. Although equivalent nonuniform MMC tolerances can be derived, they will not necessarily obtain the desired control in the projected tolerance area and are too complex to have any significant value. Hence, critical size conversions are not recommended.

PERPENDICULARITY TOLERANCE

Although perpendicularity of both clearance and fastener mounting holes is controlled by position tolerance (see Figures 8-4 and 8-6) further restriction of the perpendicularity variation can be made. While such restrictions are not needed for ordinary fastener applications, they may be required for special conditions. Consider the application shown in Figure 9-10. Here the bolt head must remain flat against the washer seal to hold hydraulic pressure within the assembly. Perpendicularity tolerance is applied to the mounting hole to attain this end.

For any application where perpendicularity is critical, it may be restricted by specifying a smaller perpendicularity tolerance than the related position tolerance. The value of the perpendicularity tolerance is given as a total value and represents the full diameter of the perpendicularity tolerance zone. The depth of the perpendicularity tolerance zone is coincident with the position tolerance for both conventional and projected tolerances. The perpendicularity tolerance zone, being lower in hierarchy, is independent of the position toler-

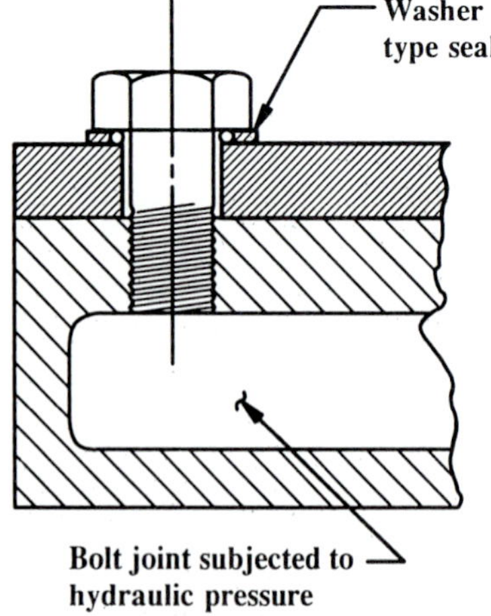

Figure 9-10. Typical mounting hole application requiring restricted perpendicularity

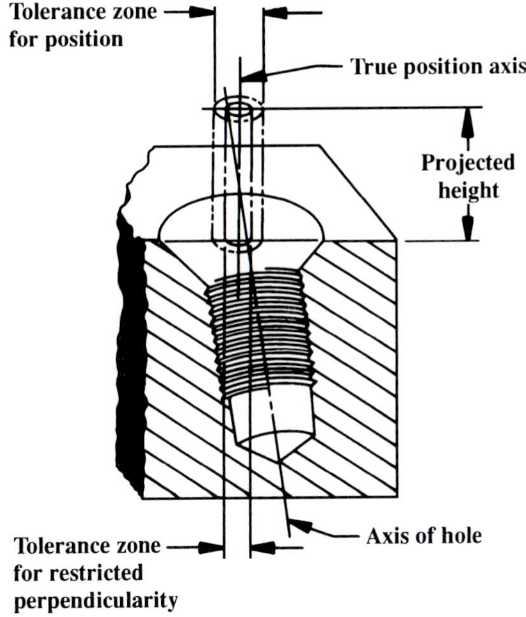

Figure 9-11. Tolerance geometry for restricted perpendicularity of a mounting hole

ance and may move any where within its limits. Figure 9-11 illustrates the restriction of perpendicularity for a fastener mounting hole. The same restrictions may be applied to clearance holes but are seldom required.

COMPARISON TO OTHER SYSTEMS

Having shown that position tolerances are based upon the functional requirements of fastener assembly, let's look briefly at conventionally tolerance parts. Figure 9-12a illustrates a conventional coordinate dimensional pattern with bilateral tolerances. Two of the many possible interpretations are shown in Figures 9-12b and 9-12c.

Unequal Tolerance

If, as in Figure 9-12b, the hole at A is considered as a datum, the resulting tolerance zones are a vertical line, a square, and a horizontal line, following the holes in a clockwise manner. As each hole has the same available clearance as any other, tolerance zones of differing shape are not logical. Further, no tolerance is allowed for hole A.

Equal-Square Tolerance

One interpretation, commonly suggested, is that the tolerance be taken as equal squares at each hole center, as shown in Figure 9-12c. This has the obvious advantage of allowing an equal tolerance for each hole. However, if we contrast these tolerance zones with those

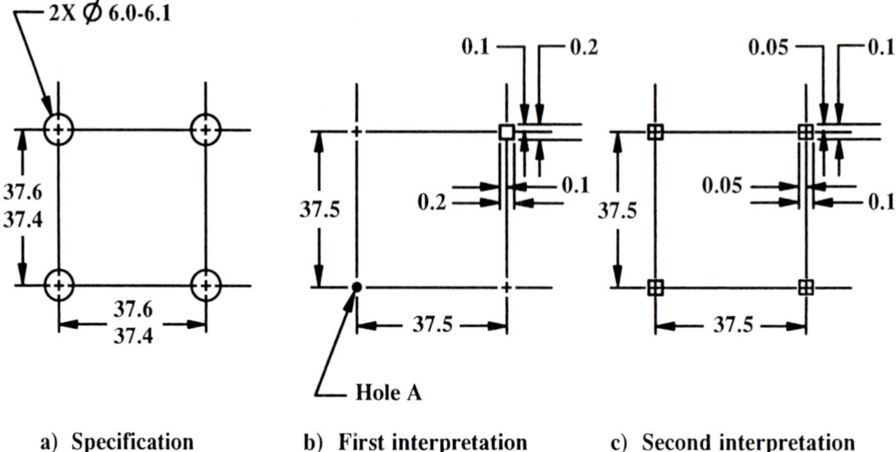

Figure 9-12. Interpretations of coordinate dimensional tolerances

developed in the foregoing section, many inadequacies are apparent. The most significant is that the shape of the tolerance zones is not related to the shape of the feature as functional considerations would require. Further, there is nothing in these specifications which indicates the extent of the tolerance zones along the axis of the holes (i.e., projected or within the hole). The concept of critical size and the attendant increase in position tolerance when the holes are not critically sized is completely lost. Therefore, due to the ambiguities, the lack of functional description, and the many other inadequacies of conventional tolerancing, only position tolerance specifications should be used for mating hole patterns.

The Increased Tolerance Fallacy

While contrasting alternate systems, one other point should be examined. The claim is often made that the use of position tolerance increases the allowable or usable tolerance. Obviously, the critical size concept may do just that. However, the substitution of a round tolerance zone for a square tolerance zone is the principal basis for most such claims. Although not frequently stressed, such an argument is valid only when round features are used.

If two tolerance zones are developed so that the same limiting position (such as Point A in Figure 9-13) is obtained, the round tolerance zone obviously allows greater position tolerance. This is represented by the shaded area which totals about 57% of the area of the inscribed square tolerance zone. The fallacy of claiming increased tolerance is that while exact tolerances can be readily calculated for position tolerance, many conventionally toleranced parts have values which at best are estimates. Therefore, saying that one system allows a certain percentage more or less tolerance is not completely realistic as it assumes that the values to be compared are accurate.

If, however, tolerances were calculated by position tolerance methods and then converted to equivalent coordinate tolerance zones, the suggested differences would be true.

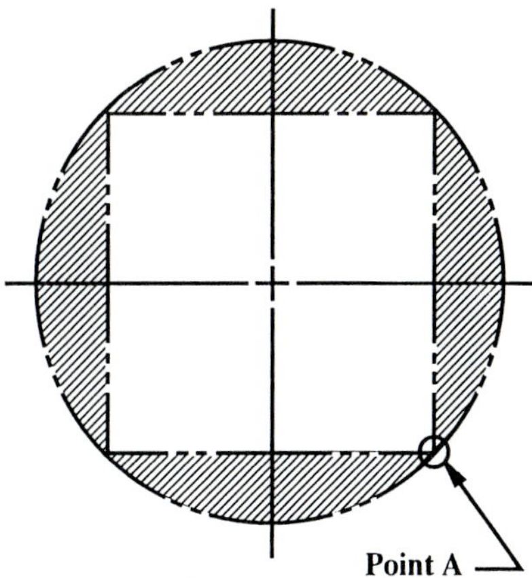

Figure 9-13. Comparison of round position tolerance zone with square coordinate tolerance zone

This may be necessary for specific uses and is discussed further in Chapter 11 on position tolerance conversions.

DIMENSIONING

It must be clearly understood that position tolerances never apply to a dimension. Instead, they define only the allowable variation in location of a feature from its true position. Dimensions locating such features define only the nominal or true position of the feature. Such dimensions are untoleranced and have no bearing on the variation that is allowed from the defined true position. Then any dimensional system, including cumulative or chain dimensions, and equal spacing notes, may be used since tolerances cannot *stack-up*. At the other end of the spectrum are CAD systems where the locations are in a math data base and no dimensions exist. These too are acceptable, again, as the tolerance applies to the location, independent of the manner in which its nominal value is defined.

It is recommended that hole patterns in mating parts be dimensioned in the same manner (see Figures 9-14a and 9-14b) in order to avoid small changes in nominal location and so that the mating hole patterns can be easily recognized.

SAMPLE SPECIFICATIONS

Figure 9-15 illustrates a typical assembly and the detailed tolerance specifications. Notes are explanatory material for the reader and tolerance zones are shown for illustrative purposes and are not normally a part of the specification.

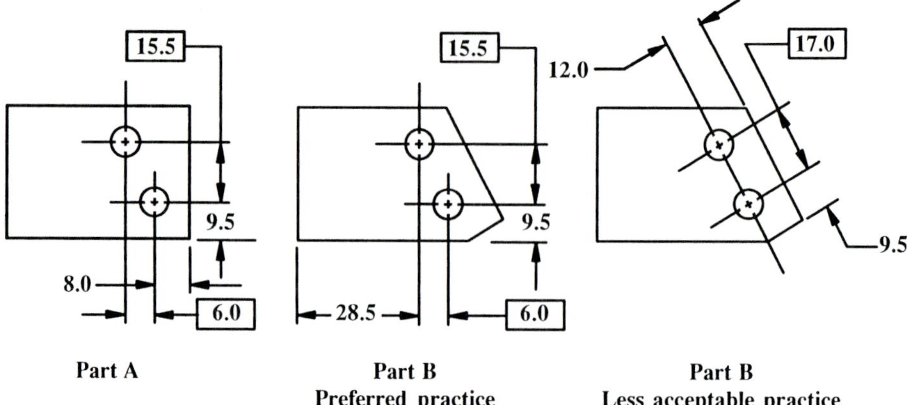

Figure 9-14. Preferred dimensioning practice. Hole patterns in mating parts should be dimensioned in the same manner

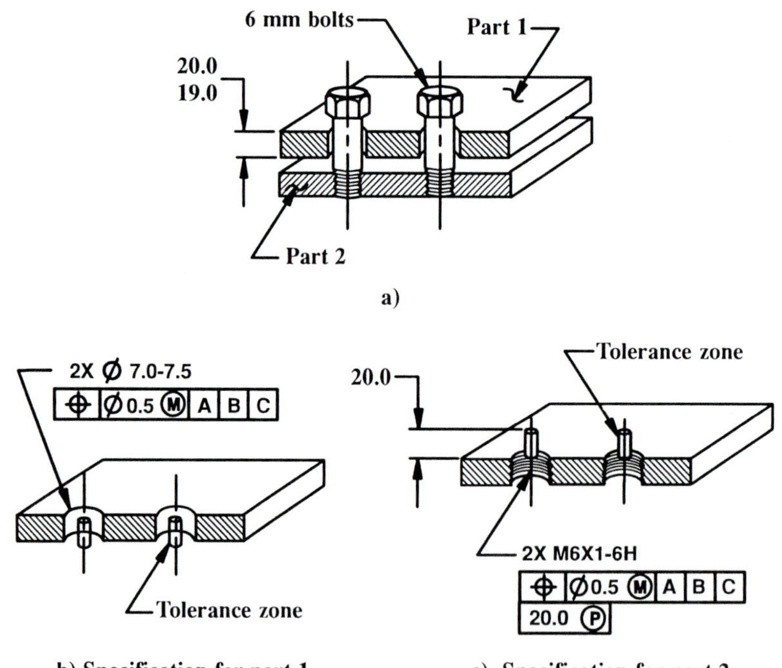

Figure 9-15. Tolerance specifications for a typical assembly

If the fasteners in Figure 9-15 were studs with a maximum height of 35mm, the specification for Part 2 would read:

2X M6X1 - 6H

⌖	Ø 0.5 Ⓜ	A	B	C
20.0 Ⓟ 35.0 Ⓕ				

A specification for dowels press fit into Part 2 would read:

2X Ø 5.88 - 5.98

⌖	Ø 0.1 Ⓢ	A	B	C
20.0 Ⓟ				

In similar fashion, if the tapped holes were replaced with weld nuts, the pilot hole specification would read:

5X Ø 10.3 - 10.6

⌖	Ø 0.5 Ⓛ	A	B	C

Figures 9-16 and 9-17 illustrate the use of tolerance specifications with base-line dimensioning and equal spacing notes, respectively. Figure 9-18 shows the use of supplemental notes to define the extent of a pattern of functionally related holes. Figure 9-19 shows alternate methods for designating counterbores, while the specification of Figure 9-20 delineates a position tolerance with a restricted perpendicularity requirement.

NONPARALLEL FEATURES

Although feature patterns usually consist of features with parallel axes, there is no reason to restrict the concept of a pattern to this convention. If a group of features function as one unit, regardless of differences in size, or plane of location, they constitute a pattern. Be-

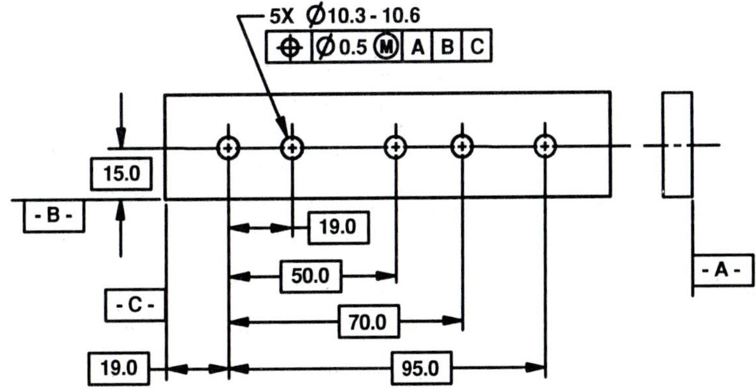

Figure 9-16. Position tolerance specification used with base line dimensioning

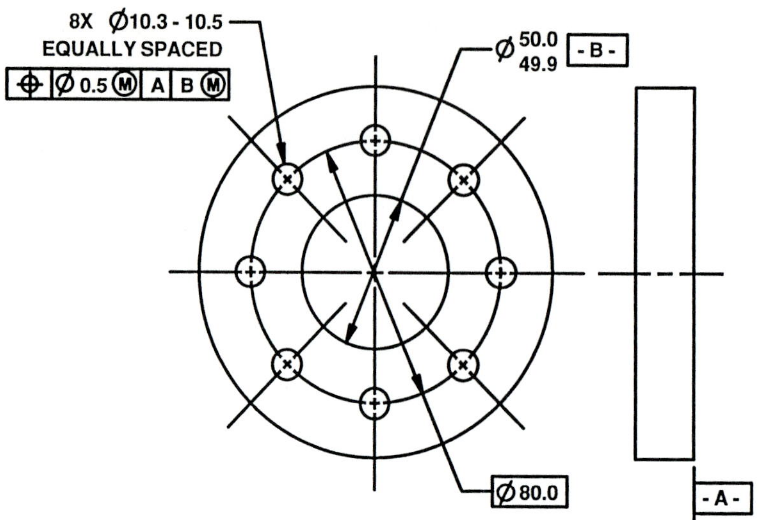

Figure 9-17. Position tolerance specification used with equal spacing note

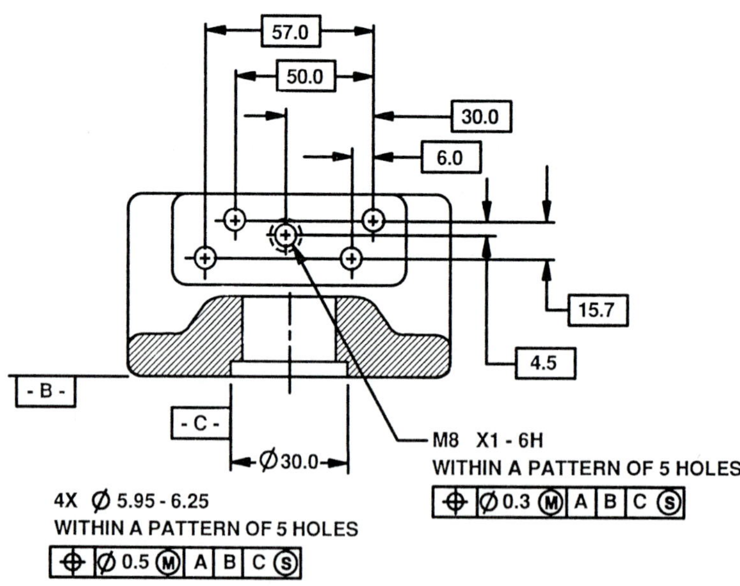

Figure 9-18. Definition of the extent of a pattern of holes by the use of supplementary notes

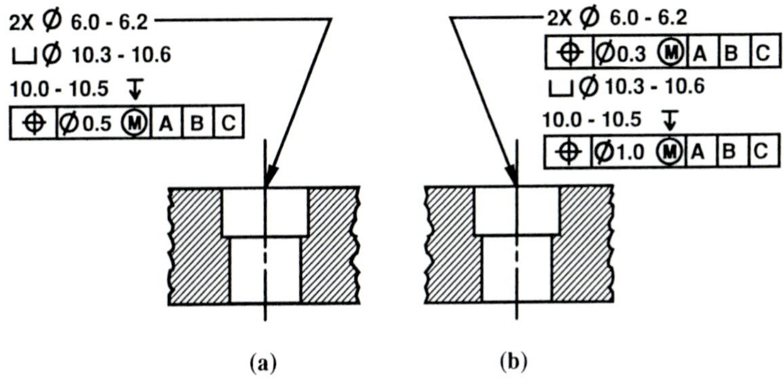

Figure 9-19. Alternate methods of designating position tolerances for counterbored holes

cause position tolerances can describe the location of any functioning feature, they can be applied within a nonparallel feature pattern equally well. Then, as in any other case, the position tolerances are constructed about the true position axes of the features, either within their depth, or projecting above the interface surface, depending upon the type of feature. Figure 9-21 illustrates a common type of nonparallel hole pattern and its related position tolerance specification.

NONUNIFORM TOLERANCE

Position tolerance standards recognize that, where it is difficult to maintain uniform tolerance control, it may be desirable to control the location of a feature more accurately at on end than at the other. This can be accomplished by specifying a larger tolerance at the less critical end of the feature as shown for the hole in Figure 9-22.

If we reexamine Figure 8-6, we see that most functional installations will use uniform tolerances. Then, for a nonuniform tolerance application to be valid, it must be other than a conventional fastener application. In line with this reasoning, nonuniform tolerances have been applied to such applications as long drilled oil passages in engine blocks where the holes must be accurately aligned at the gasket face and cannot be located with the same

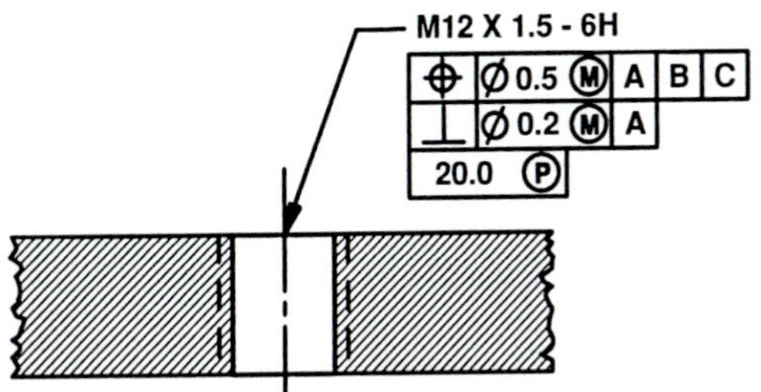

Figure 9-20. Position tolerance with a restricted perpendicularity requirement

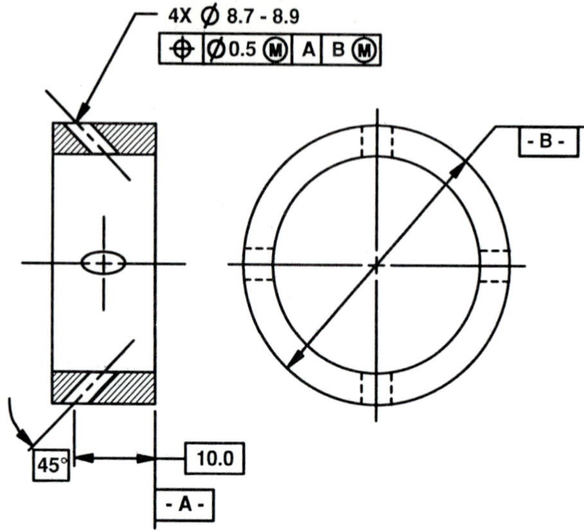

Figure 9-21. Position tolerance applied to a pattern of non-parallel holes

accuracy along their full length. Figure 9-23 typifies the specification for such an application.

GENERAL TOLERANCE SPECIFICATIONS

Since most printed drawing forms used in industry utilize a general tolerance specification, its relation to positionally tolerance features should be examined. Such general tolerance specifications always include the statement *unless otherwise specified.* Then dimensions

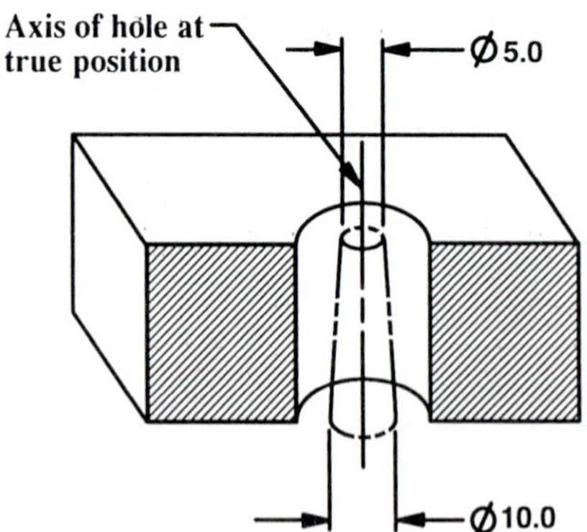

Figure 9-22. Tolerance geometry for non-uniform position tolerance

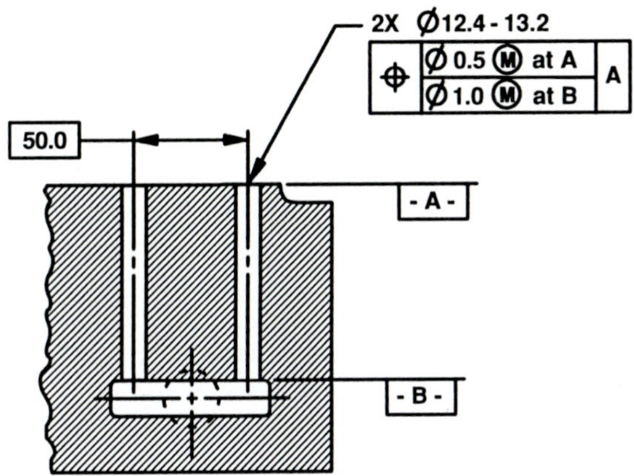

Figure 9-23. Position tolerance specification for special application requiring non-uniform position tolerance

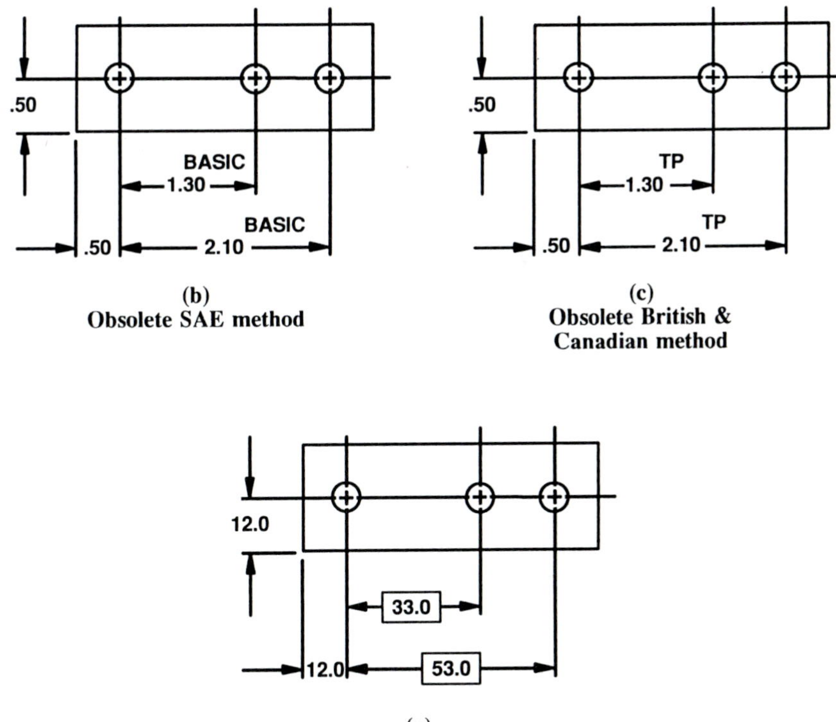

Figure 9-24. Dimensions exempt from general tolerance

defining the true position of a positionally toleranced feature are clearly exempt because the feature location is controlled by the position tolerance. As noted in Chapter 5, ANSI-Y14.5M defines the use of a rectangular frame around a dimension to denote a *basic dimension* (see Figure 9-24a). Figures 9-24b and 9-24c show obsolete notations that may be encountered occasionally on old drawings.

SUMMARY

The boundary, or tolerance zone, for a fastener mounting hole requires projected height or projected height and fastener height values in order to define its extent.

Size and location tolerance are related and can be formulated as a function of the critical feature sizes. Critical sizes are also related to function as shown below:

Critical Size	*Function*
MMC	Clearance
LMC	Location
RFS	Mounting

For special applications, feature perpendicularity may be restricted beyond the control provided by the position tolerance. Dimensioning of positionally toleranced features can be done by almost any convenient method because position tolerances do not *stack-up*. Dimensions defining true positions are identified as basic to indicate that general tolerances do not apply.

Chapter 10

Applications of Position Tolerance

This chapter develops equations for determining the values of critical tolerance for position tolerance specifications. It also discusses common fastener installations which are then used to illustrate the variety of position tolerance applications. The examples given show the many approaches to problem solving. With a thorough understanding of the methods developed in this chapter, you will be able to cope with unusual problems by employing these same principles.

FLOATING FASTENERS

A *floating fastener* is defined as an application in which the fastener passes through clearance holes in two or more parts.

To determine the effect of variation in hole location, let's examine the most general case represented by two parts with clearance holes and a bolt. These parts are shown in Figure 10-1a. In this figure, the clearance between the bolt and the hole in Part 1 is not equal to the clearance at the hole in Part 2. In this instance, the difference is the result of unequal hole sizes, but a stepped diameter fastener body would create a similar situation. Assume that the body of the fastener remains perpendicular with the surfaces of Parts 1 and 2 at all times. Since this application depends upon clearance to assemble, the holes are critical at their MMC size.

In analyzing Figure 10-1a, we see that the total tolerance is equal to two times the movements which close the clearances A and B to bring the edges of the holes in both parts against the fastener (see Figure 10-1b). Therefore, the total tolerance equals two times ($A + B$). But $A + B$ is the average diametral clearance to the bolt in Parts 1 and 2.

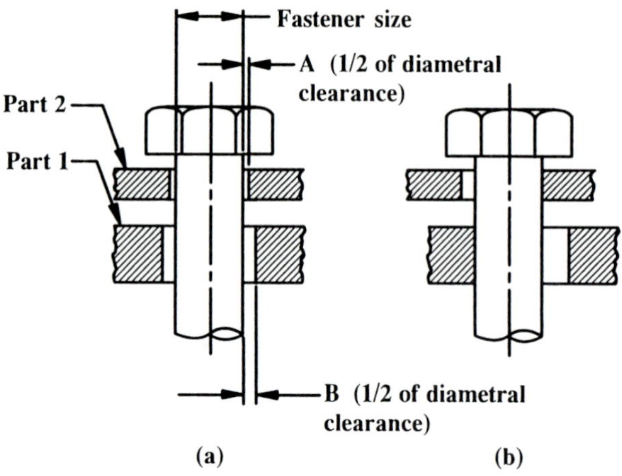

Figure 10-1. Floating fastener application

$$T_T = 2(A + B) = CL_1 + CL_2$$

$$T_T = 2\ CL_{AV} \qquad (10\text{-}1)$$

where: T_T = Total position tolerance for two parts
CL_{AV} = Average of two clearances

When manufacturing processes are such that the greatest accumulation of error occurs in one part, unbalanced tolerances are often desirable. For example, a part with fixture drilled holes might require less tolerance than a mating part with punched holes where the punches are positioned by a setup. Judicious use of position tolerance can provide a large tolerance where it will be of greatest benefit by dividing the total tolerance, T_T, between the two parts in any desired proportion. If the processes are similar, and balanced tolerances are desired, the total tolerance may be divided equally. This may be calculated directly from the following formula:

$$T_B = CL_{AV} \qquad (10\text{-}2)$$

where: T_B = Balance position tolerance.

While the necessity of using different clearances occurs for special fasteners, and in situations such as adapting a purchased or existing component with established hole size and Position tolerance to a new design, the use of identical hole sizes is more common. For such applications the average clearance becomes simply the clearance, *CL*, at either hole. The tolerance formulas simplify to:

$$T_T = 2CL \qquad (10\text{-}3)$$

$$T_B = CL \qquad (10\text{-}4)$$

Alternate Equations

Since clearances can be expressed as the difference of the hole size, H, and the fastener size, F, the tolerance equations can be expressed in these terms:

$$\text{From:} \quad CL = H - F$$

$$T_T = (H_1 + H_2) - (F_1 + F_2)$$

$$= H_1 + H_2 - 2F, \text{ if } F_1 = F_2 \qquad (10\text{-}1a)$$

$$= 2H - (F_1 + F_2), \text{ if } H_1 = H_2 \qquad (10\text{-}1b)$$

$$T_B = \frac{(H_1 + H_2) - (F_1 + F_2)}{2}$$

$$= \frac{H_1 + H_2}{2} - F, \text{ if } F_1 = F_2 \qquad (10\text{-}2a)$$

$$= H - \frac{F_1 + F_2}{2}, \text{ if } H_1 = H_2 \qquad (10\text{-}2b)$$

$$T_T = 2(H - F) \qquad (10\text{-}3a)$$

$$T_B = H - F \qquad (10\text{-}4a)$$

$$\left.\right\} \text{ If } H_1 = H_2, \text{ and } F_1 = F_2$$

These relations may also be used to calculate hole sizes where a known or desired position tolerance is specified. This procedure is shown in several of the examples.

Multiple Component Assemblies

If more than two parts are assembled in a floating fastener application, tolerances assigned must be such that any two parts will mate. For such an assembly, calculate any two parts and then calculate each remaining part to mate with each of the preceding parts.

Stationary Fastener Method

If it is assumed that the fastener does not move from its true position, multiple part assemblies can be calculated by a simplified method. In this procedure, each hole is toleranced to mate with the fastener at its fixed position according to its size. Examination of Figure 10-1a discloses that each hole may move by one-half of its MMC clearance, indicating that Equations 10-4 and 10-4a may be used.

There are two disadvantages inherent in this approach. First, if unbalanced tolerances are desired, different hole sizes must be used. Second, calculations for desired position tolerance usually lead to odd or nonstandard holes sizes, since the tolerances are not easily unbalanced. Additionally, this method must not be used when adding new parts to mate with existing designs because the limitations imposed by this technique may not have been observed in the original design calculations.

Example 10-1

Given a 10mm bolt[1] with 10.5 – 10.9 diameter clearance holes, calculate an unbalanced tolerance arrangement.

Solution From Equation 10-3a:

$$T_T = 2(10.5 - 10.0) = 1.0$$

The tolerance may be distributed between the two parts in any manner just as long as the sum of 1.0 is maintained (e.g., 0.7 on one part and 0.3 on the other).

Example 10-2

Calculate balanced tolerances for the preceding problem.

Solution From Equation 10-4a:

$$T_B = 10.5 - 10.0 = 0.5$$

Example 10-3

An existing component has 10.7 – 11.0 diameter clearance holes with a specified position tolerance of ∅ 1.0. Design a mating part to accept a 10mm diameter bolt. position tolerance for the new part is to be ∅ 0.7.

Solution

$$\begin{aligned} T_T &= 1.0 + 0.7 = 1.7 \\ T_T &= H_1 + H_2 - 2F \text{ (Equation 10-1a)} \\ 1.7 &= 10.7 + H_2 - 2 \times 10.0 \\ H_2 &= 20.0 - 10.7 + 1.7 = 11.0 \end{aligned}$$

Example 10-4

Three parts are to be fastened together as shown by the cross section through one bolt.

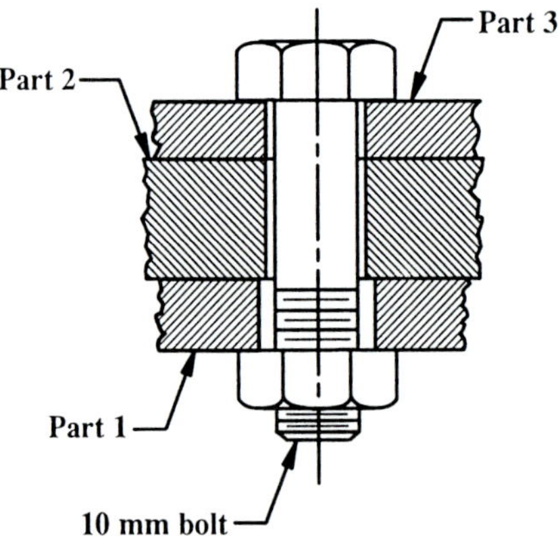

[1]The nominal size of fasteners is often the MMC size.

Part 1 is a stamping and may not use a position tolerance less than ∅ 0.8. Parts 2 and 3 are drilled and a standard drill size is to be used. Assume a drilled hole size tolerance of −0.05. If a minimum position tolerance of ∅ 0.3 is acceptable for parts 2 and 3, calculate all of the hole sizes.

Solution Balanced tolerances may be used for Parts 2 and 3 since they are made by identical processes.

Applying Equation 10-4a:
$$T_B = H - F$$
$$0.3 = H - 10$$
$$H = 10.3$$

Nominal hole size = 10.3 + 0.05 + 10.35

Closest metric drill size = 10.4

Minimum hole size = 10.4 − 0.05 = 10.35

For Part 1 to mate with Part 2 or 3:

$$T_T = 0.8 + 0.3 = 1.1$$

Applying Equation 10-1a:
$$T_T = (H_1 + H_2) - 2F$$
$$1.1 = (10.35 + H_1) - 2 \times 10.0$$
$$H_1 = 20.0 - 10.35 + 1.1 = 10.75$$

Example 10-5

Solve the preceding problem by the stationary fastener method.

Solution

For Parts 2 and 3: $H = 10.35$ as above

For Part 1:
$$T_B = H - F$$
$$0.8 = H_1 - 10.0$$
$$H_1 = 10.8$$

Notice that both methods give the same critical size for Parts 2 and 3 due to the balanced tolerance requirement stated in Example 10-4. When the hole size is increased to a standard drill size, however, the hole size of Part 1 is reduced only when the more general method (Example 10-4) is used. With the stationary fastener method, the standard hole size selected should be used to calculate an increased position tolerance for Parts 2 and 3.

Example 10-6

Calculate the position tolerance for a fourth part to mate with the parts of Example 10-4. Use an MMC hole size of 11.4.

Solution Applying Equation 10-1a to Parts 1 and 4:

$$T_T = H_1 + H_4 - 2F = (10.75 + 11.4) - 2 \times 10.0$$
$$T_T = 22.15 - 20.0 = 2.15$$
$$T_4 = T_T - T_1 = 2.15 - 0.8 = 1.35$$

Applying Equation 10-1a to Parts 2 (or 3) and 4:

$$T_T = (10.35 + 11.4) - 2 \times 10.0$$
$$T_T = 21.75 - 20.0 = 1.75$$
$$T_4 = T_T - T_2 = 1.75 - 0.3 = 1.45$$

Selecting the smallest value: $T_4 = 1.35$

FIXED FASTENERS

A *fixed fastener* is defined as an application in which the fastener is held in a mounting hole in one part and passes through clearance holes in subsequent parts.

To determine the effect of variation in hole location for this type of fastener, let us examine a typical application consisting of a bolt mounted into a tapped hole (see Figure 10-2a). As either the clearance or mounting hole moves, the effect is to close the clearance between the bolt and the edge of the clearance hole, as shown in Figure 10-2b. From this movement, we see that the total tolerance is equal to the diametral clearance or Equation 10-5. If unbalanced tolerances are desired, the total tolerance, T_T, may be divided between the two parts in any proportion. If balanced tolerances are desired, the total tolerance is divided equally, yielding Equation 10-6. If the fixed fastener application is compared to the floating fastener application, the major difference to be noted is that one clearance hole must now absorb all displacements. Hence, a fixed fastener tolerance is numerically one-half the value of a similar floating fastener application, as you might anticipate.

$$T_T = CL, \text{ and} \tag{10-5}$$

$$T_B = \frac{CL}{2} \tag{10-6}$$

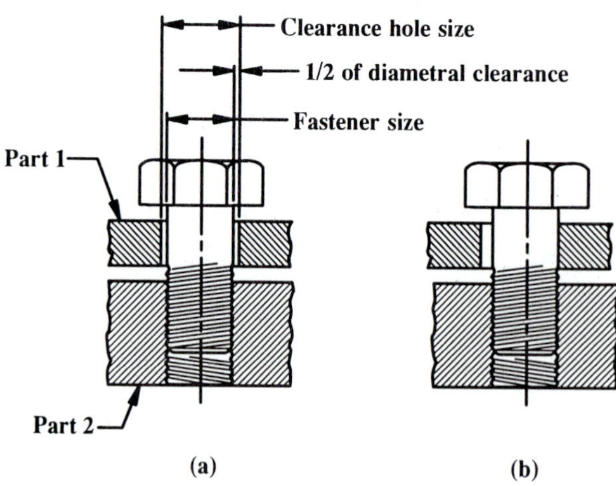

Figure 10-2. Fixed fastener application

Or in terms of hole size (H) and fastener size (F):

$$T_T = H - F, \text{ and} \qquad (10\text{-}5a)$$

$$T_B = \frac{H - F}{2} \qquad (10\text{-}6a)$$

In examining Figure 10-2, note that the clearance hole is an MMC application, as is any clearance hole. The tapped hole is a mounting hole and must be controlled by use of projected and fastener heights. In some cases, such as a dowel press fit into the mounting hole, an RFS application is required.

Multiple Component Assemblies

If more than two parts are assembled in a fixed fastener application, each part containing clearance holes must be calculated to mate with the part containing the fastener mounting hole.

Example 10-7

Calculate balanced and unbalanced tolerances for a 10mm diameter bolt mounted in a tapped hole and passing through a 11.0 – 11.5 diameter clearance hole.

Solution Using Equation 10-5a for total tolerance:

$$T_T = 11.0 - 10.0 = 1.0$$

Tolerances may be distributed as desired (e.g., 0.4 for the clearance hole and 0.6 for the tapped hole). A projected height must be specified for the tapped hole.

Using Equation 10-6a for balanced tolerances:

$$T_B = \frac{11.0 - 10.0}{2} = 0.5$$

Example 10-8

Calculate the position tolerance for a third part with a minimum clearance hole size of 11.6 diameter to mate with the parts of Example 10-7. Assume that the balanced tolerance was used.

Solution From Equation 10-5a:

$$T_T = 11.6 - 10.0 = 1.6$$
$$T_2 = T_T - T_1 = 1.6 - 0.5 = 1.1$$

Fasteners with Eccentricity

If a stepped fastener, such as the shoulder bolt shown in Figure 10-3a, has an allowable eccentricity, E, it will affect the tolerance-clearance relations.

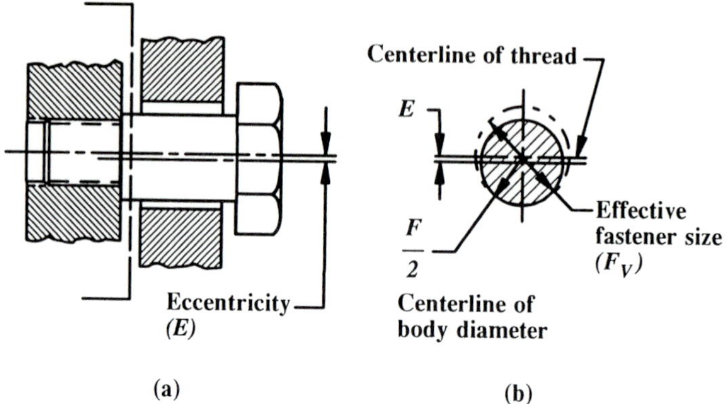

Figure 10-3. Stepped fixed fastener with allowable eccentricity

A radius swung from the center of the mounting hole to the extreme point of eccentricity (see Figure 10-3b) will define a virtual fastener size, F_V, which includes the allowable eccentricity. Then in terms of the MMC fastener size, F, and allowable eccentricity, E:

$$F_V = 2\left[\frac{F}{2} + E\right]$$

$$F_V = F + 2E, \text{ or} \tag{10-7}$$

$$F_V = F + C \tag{10-8}$$

where C is a Concentricity (or Position Tolerance) Limit (C = 2E)

In any problem where an eccentricity can occur, the fastener size may be replaced with the virtual fastener size and any of the standard tolerance equations can then be used. If a virtual size is used for a floating fastener application, note that the clearance in only one of the two parts can be affected by an eccentricity. Moderate values of eccentricity are often ignored for floating fastener applications consisting of only two parts as the fastener can be turned to its most advantageous position. Since eccentricities are frequently encountered with fixed fastener applications, the tolerance-clearance equations can be modified to permit direct calculation by substituting $F = F_V$. Then:

$$T_T = H - (F + C) \tag{10-5b}$$

$$T_B = \frac{H - (F + C)}{2} \tag{10-6b}$$

Positive Clearance at Assembly

When a positive clearance is required at assembly, the virtual fastener size concept can be used by substituting the value of the required clearance for the concentricity, C. This may be applied to floating, fixed, or piloted applications.

Example 10-9

Referring to Figure 10-3, use the following parameters: MMC hole diameter = 11.3, MMC body diameter = 10mm, concentricity thread to body = .25 TIR. Calculate an acceptable balanced position tolerance.

Solution Applying Equation 10-6b:

$$T_B = \frac{11.3 - (10.0 + .25)}{2} = \frac{1.05}{2} = 0.52$$

and the value is rounded off to 0.5.

Remember that a projected height must be specified for the mounting hole.

MMC CONCENTRICITY FOR FASTENERS

When we considered the effect of eccentricity on a fastener, we arrived at the concept of virtual fastener size; this virtual size being a function of the eccentricity, E or C, and the MMC fastener size, F. If the fastener size is decreased, while the allowable eccentricity is maintained, the fastener surface moves away from the established boundary which we call virtual size, as in Figure 10-4.

Since the virtual size is used in the position tolerance calculations, the space within its boundary is always available for errors of fastener form. Then, rather than to specify a fixed value of eccentricity, critical size specifications are used. In the past, this resulted in an MMC concentricity specification, which was defined so that the specified value of concentricity applied at an MMC critical size. This allows the virtual size to be maintained and the concentricity to increase when the fastener size decreases. As noted in Chapter 7, concentricity tolerances are now defined as RFS controls. Hence, MMC critical position tolerances are now the best tool to define such a relationship. An example is shown in Figure 10-5.

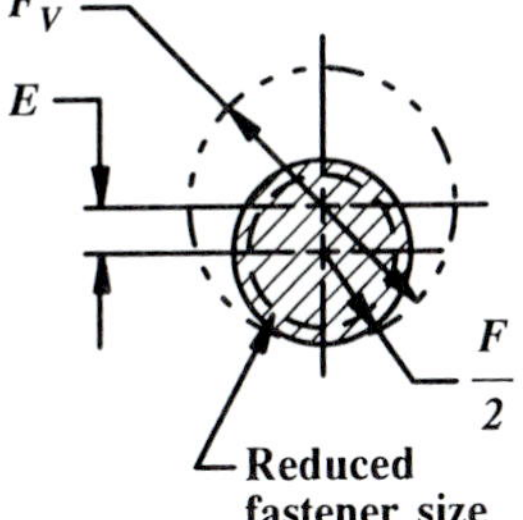

Figure 10-4. MMC concentricity application. Use of this specification for a stepped fastener results in a constant virtual size

COUNTERBORES

Although they are frequently encountered, concentricity specifications for counterbores are inappropriate because a functional description would allow a clearance hole and a counter bore to move in opposite directions, as if they were in separate parts. Position tolerances

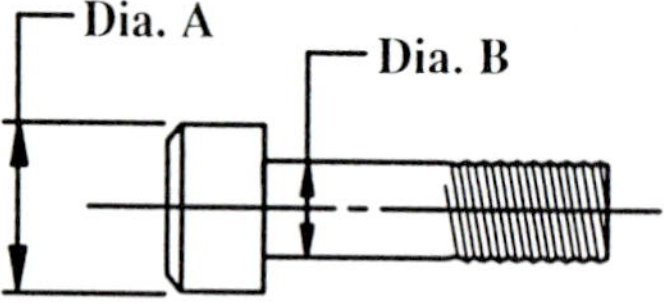

FOR A and B

⌖	Ø 0.1 Ⓜ	THREAD P.D

Figure 10-5. Typical MMC concentricity specification

can be applied to counterbored holes by using the same techniques used for clearance holes in more than two parts. The method to be used must be selected to correspond with the application: that is, either fixed or floating fasteners may be encountered. To account for allowable eccentricities, the concept of virtual size must be used.

Example 10-10

Data for the figure shown is: bolt body diameter = 6.0, bolt head = 10.0, concentricity of body to head = 0.1 TIR. In both parts, the MMC hole size = 6.4mm diameter and the MMC counterbore size = 10.7 diameter. Develop balanced tolerances for the holes and the resulting position tolerance for the counterbore.

Solution Solving for the clearance hole position tolerance with Equation 10-4a:

$$T_B = H - F = 6.4 - 6.0 = 0.4$$

Solving for the position tolerance at the counterbores with Equations 10-8 and 10-1a:

$$\begin{aligned} F_V &= F + C = 10.0 + 0.1 = 10.1 \\ T_T &= (H_1 + H_2) - (F_1 + F_2) \\ T_T &= (10.7 + 6.4) - (10.1 + 6.0) = 1.0 \\ T_{CB} &= T_T - T_B = 1.0 - 0.4 = 0.6 \end{aligned}$$

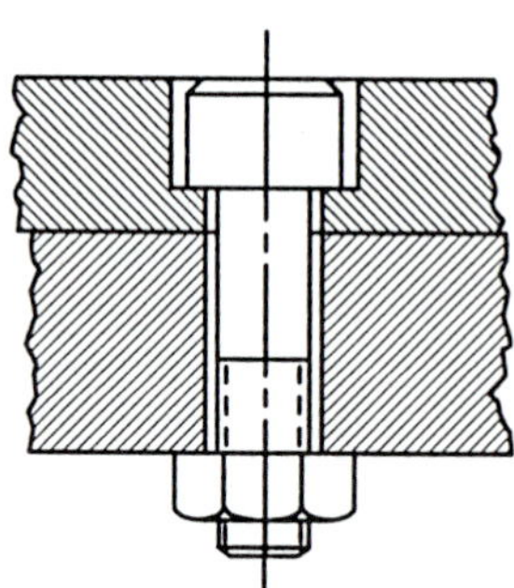

Example 10-11

Using data from Example 10-10, calculate position tolerances for the fixed fastener shown below. Concentricity of thread to body = 0.1 TIR, of thread to head = 0.2 TIR. Use 0.2 position tolerance at the tapped holes.

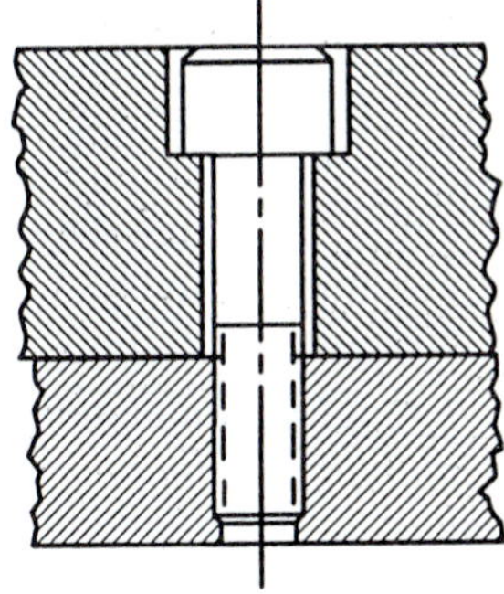

Solution Using Equation 10-5b to find the clearance hole tolerance:

$$T_T = H - (F + C) = 6.4 - (6.0 + 0.1) = 0.3$$
$$T_2 = T_T - T_1 = 0.3 - 0.2 = 0.1$$

Solving for the counterbore tolerance in the same fashion:

$$T_T = 10.7 - (10. + 0.2) = 0.5$$
$$T_{CB} = T_T - T_1 = 0.5 - 0.2 = 0.3$$

Example 10-12

Referring to the Figure in Example 10-11, assume that the bolt size is 6mm and that position tolerance for the tapped holes is 0.25. Calculate a recommended clearance hole size for minimum position tolerances of 0.15 at the clearance hole and counterbore.

Solution Manufactures of socket head screws have established gage sizes for controlling eccentricity in their products. These gage sizes are really virtual sizes and can be used directly in position tolerance calculations.

Given that F_V Body = 6.1, and

F_V Head = 10.2

Solving for clearance hole size using Equation 10-5a:

$$T_T = H - F$$
$$\text{but } T_T = 0.25 + 0.15 = 0.4$$
$$0.4 = H - 6.1$$
$$H = 6.5$$

Typical Specification

⌀ 6.5 - 7.0

⌖	⌀ 0.15 Ⓜ	DATUM

Solving for the counterbore size using the same approach:

$$T_T = 0.4 \text{ as above}$$
$$0.4 = H - 10.2$$
$$H = 10.6$$

Typical Specification

⌴⌀ 10.6 - 11.1 10.0 ↧

⌖	⌀ 0.15 Ⓜ	DATUM

CARRIAGE BOLTS

The *carriage bolt,* as shown in Figure 10-6, is usually mounted through a round hole at the body section and a square hole at the neck section and is essentially a floating fastener application. While the round hole retains the familiar round boundary and tolerance zone, Figure 10-7 shows that the cross section of the square hole, subject to the limitation of a square boundary, or virtual size, will trace out a tolerance zone which has a square cross section. The three-dimensional tolerance zone for a square hole is a rectangular prism with a square base which coincides with the depth of the hole. As in previous cases, either the three-dimensional boundary or tolerance zone controls both positional and perpendicularity variations. Due to the differing sections, it is observed that the formulas based upon average clearance will have to be used to calculate tolerance values.

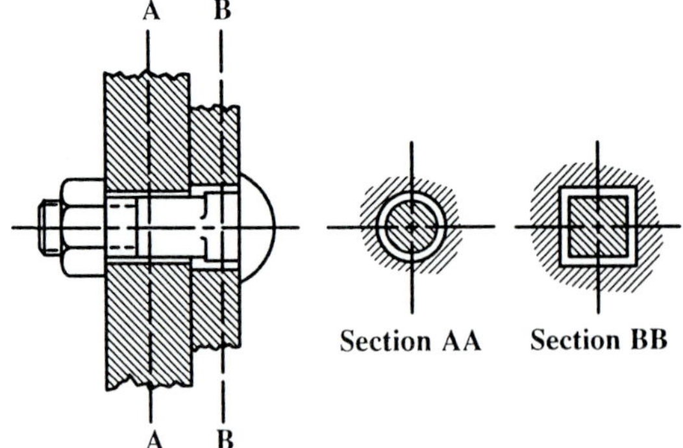

Figure 10-6. Differing cross sections through a carriage bolt and its clearance holes

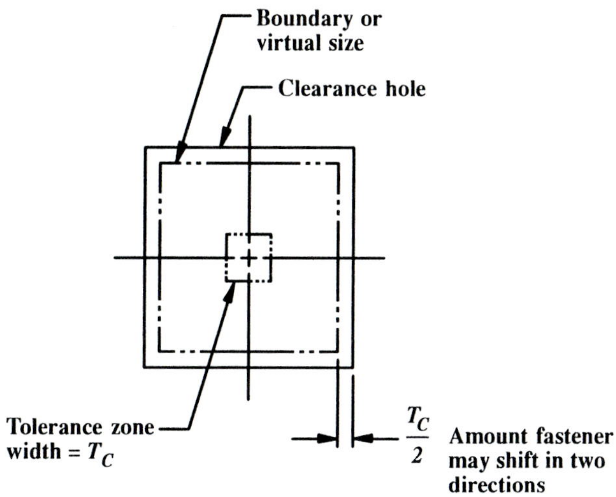

Figure 10-7. Boundary and tolerance zone geometry for a square hole

Relation between Round and Square Cross Sections

To determine if the standard floating fastener formulas are valid, let's consider the effect of relative movement of the square and round holes within the limits imposed by their respective tolerance zones. In the first analysis, assume that the position tolerance at each hole is equal to the assembly clearance at that hole (Stationary Fastener Method). Then for this case, each hole may move up to the bolt, but cannot displace the bolt in the other hole due to the balance between tolerance and assembly clearance. Obviously, this tolerance arrangement is acceptable.

In the second analysis, assume that the position tolerance for the round hole is greater than its clearance by some amount *M*. A movement of the round hole equal to the full value of its position tolerance will cause the bolt to move in the square hole by one-half of the value of *M*. Because this value is fixed and can be applied in any direction, the square section of the bolt can be shifted against its boundary by movements parallel to the square hole edges, but not by movement in any other direction. This is illustrated in Figure 10-8. Then, as the boundary is not violated, the tolerance zones are compatible.

In the final analysis, assume that the position tolerance for the square hole is larger than the clearance across its section by some amount *M*. If the square hole is displaced by an amount equal to the full value of its position tolerance, it will cause the bolt to move in the round hole by an amount equal to one-half of the value of *M*. If the movement of the square hole is parallel to the square hole edges, it will cause the bolt to move against its boundary limit in the round hole (see Figure 10-9a). If the movement of the bolt is along the diagonal, however, the displacement of the bolt is greater than the one-half *M* value and exceeds the clearances upon which the position tolerances are based, resulting in possible interference as shown in Figure 10-9b.

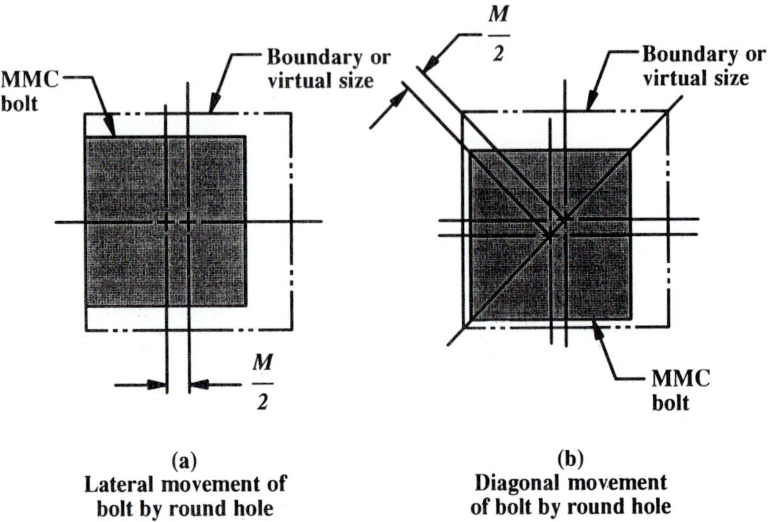

Figure 10-8. Effect of relative movement of the round hole upon the square hole in a carriage bolt application

Although the tolerance structures conflict in this case, the use of mixed square and round tolerance zones is necessary to permit functional description of the hole location requirements in the two parts. So that the standard floating fastener equations may be used for all carriage bolt applications, a correction factor will be introduced to reduce the position tolerance when the situation illustrated in Figure 10-9b is encountered. If the stationary fastener method of calculating tolerances is used, the result is as described in the first analysis and a correction factor is not necessary.

In Figure 10-10, let the lateral amount the bolt can move the square hole be represented by $M/2$ where:

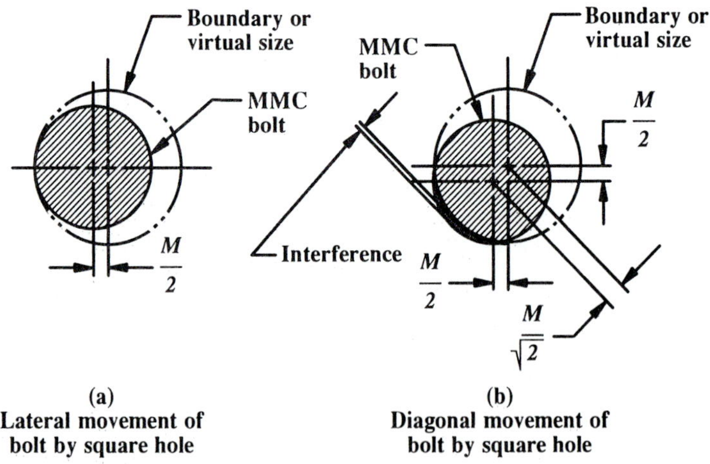

Figure 10-9. Effect of relative movement of the square hole upon the round hole in a carriage bolt application

$$M = (T_C - CL) \tag{10-9}$$

R_B = Radius of the bolt

K_R = Correction factor for round hole

K_S = Correction factor for square hole

Several important conclusions can be drawn from Figure 10-10. First, for the interference condition to exist, the factor M_S (at the square hole) must have a positive value. Hence, Equation 10-9 constitutes a test which can be used to determine if a correction factor is needed. If M is negative or zero, the calculated position tolerances are acceptable. If, however, M has a positive value, correction is necessary and may be accomplished by reducing the position tolerance value for either part by an appropriate correction factor.

Reduction of the position tolerance for the square hole causes the amount the bolt can move along the diagonal to be reduced so that the boundary of the round hole cannot be violated. In more favorable positions (other than along the diagonal), the bolt cannot move to the boundary of the round hole and a nominal clearance will exist. If the position tolerances for the round hole is reduced, its boundary size expands to include the extreme position of the bolt. In more favorable positions, as above, the bolt does not contact the boundary and a nominal clearance will exist.

Solving for K_R and K_S from Figure 10-10:

$$K_R = M\sqrt{2} - M = (\sqrt{2} - 1)M = .41M \tag{10-10}$$

$$K_S = \frac{K_R}{\sqrt{2}} = .30M \tag{10-11}$$

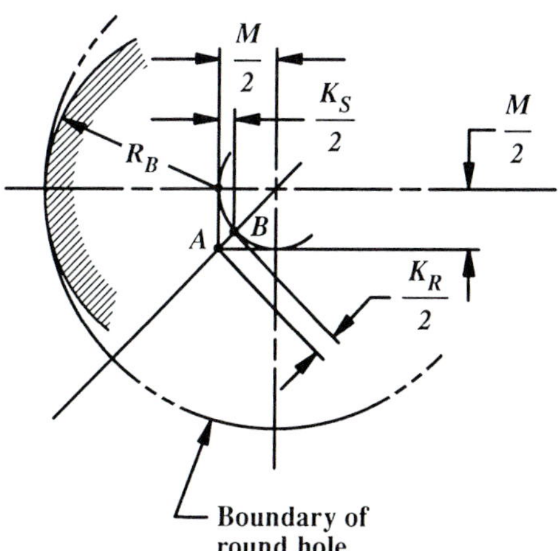

Figure 10-10. Center of bolt kept within circle of radius $M/2$. Under this condition the boundary limit of the round hole will not be violated

The values K_R and K_S permit modification of either tolerance and are useful when one tolerance is fixed and total tolerance calculations are being performed. Figure 10-11 illustrates the development of a factor, K_B, by which the common tolerance may be modified when calculating balanced tolerances.

Solving for K_B from Figure 10-11:

$$K_B + \sqrt{2}K_B = K_R = .41M$$

$$K_B = \frac{.41M}{1 + \sqrt{2}} = .18M \qquad (10\text{-}12)$$

The correction factors which are applicable to position tolerances calculated by floating fastener equations are summarized below:

$$\text{For} \quad M_S = (T_C - CL) > 0$$

$$K_R = .41M \qquad (10\text{-}10)$$

$$K_S = .30M \qquad (10\text{-}11)$$

$$K_B = .18M \qquad (10\text{-}12)$$

where:
T_C = Position tolerance at square hole
CL = MMC clearance at square hole
K_B = Correction factor for balanced tolerance
K_R = Correction factor for round hole only
K_S = Correction factor for square hole only

These correction factors can be read directly from the nomogram of Figure 10-12. For example, it the clearance is .50, and the position tolerance is .75 , connect .50 on the CL

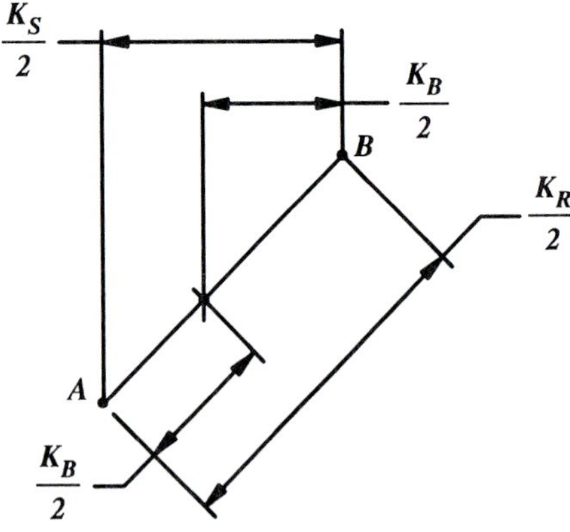

Figure 10-11. Boundary and tolerance zone geometry for a square hole

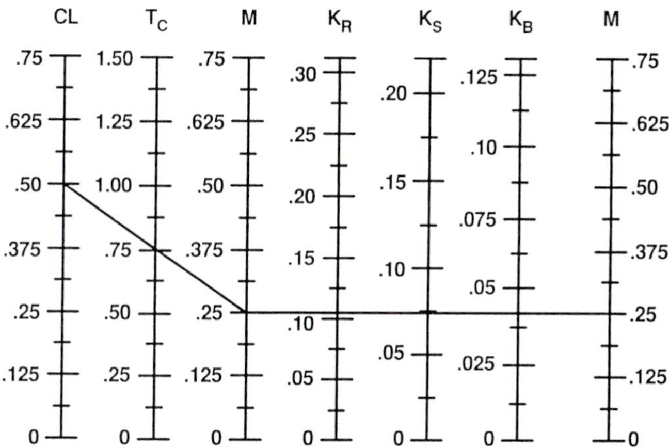

Figure 10-12. Nomogram providing correction factors for carriage bolt applications

scale with .75 on the T_C scale ; extend to M and read .25. Draw a line from .25 on the central M scale to .25 on the right-hand M scale and read the desired correction factor from: K_R = .10+, K_S = .075, or K_B = .04+.

Example 10-13

Given a 10.0 diameter bolt with a maximum neck size equal to 10.6 mounting into two parts, both containing 10.9 – 11.1 square holes. The neck of the bolt engages both parts. Solve for balanced tolerances.

Solution Since the clearance is the same at both holes, use Equation 10-4a:

$$T_B = H - F = 10.9 - 10.6 = 0.3$$

Example 10-14

In the previous example, substitute a round hole, 10.3–10.6 in diameter, for the second hole. Assume that the neck of the bolt engages only the square hole which has a position tolerance of 0.4. Find an acceptable position tolerance for the round hole.

Solution Solving for unbalance tolerances by use of Equation 10-1a:

$$\begin{aligned} T_T &= (H_1 + H_2) - (F_1 + F_2) \\ T_T &= (10.3 + 10.9) - (10.0 + 10.6) = 0.6 \\ T_2 &= 0.6 - 0.4 = 0.2 \end{aligned}$$

Calculating M_S to determine if a correction factor is needed:

$$\begin{aligned} M_S &= (T_C - CL) = 0.4 - (10.9\text{–}10.6) = 0.1 \\ K_R &= .42\text{M} = .42 \times 0.1 = 0.04 \end{aligned}$$

Tolerance for the round hole now becomes:

$$T_2 = 0.2 - 0.04 = 0.16$$

Example 10-15

Resolve Example 10-14 for balanced tolerances.

Solution Applying Equation 10-2a:

$$T_B = \frac{(H_1 + H_{2)} - (F_1 + F_{2)}}{2}$$

$$T_B = \frac{(10.3 + 10.9) - (10.0 + 10.6)}{2} = 0.3$$

$$\text{CL} = 10.9 - 10.6 = 0.3$$

$$M_S = (T_C - CL) = (0.3 - 0.3) = 0$$

Hence, correction is not required.

A typical tolerance specification for a pattern of square holes is shown in Figure 10-13.

Angular Orientation of Square Holes

The boundary of the square hole limits it angular position about its axis. This limitation results when the virtual size is constructed from the critical hole size and position tolerance (see Equation 8-1). Functionally, this limitation is required only when the neck of the bolt engages square holes in two or more parts, or when orientation of the hole is required to place the corner in a favorable position to avoid stress concentration, cracking between adjacent holes, or the like. When it is not necessary to maintain such control, a specification may be used to relax, or eliminate, this requirement. Figures 10-14 and 10-15 illustrate this technique, the former allowing small rotations of the hole, and the latter allowing complete freedom of rotation.

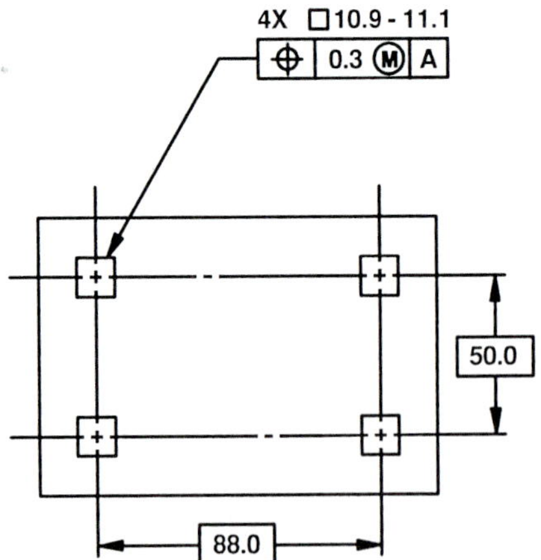

Figure 10-13. Tolerance specification for a

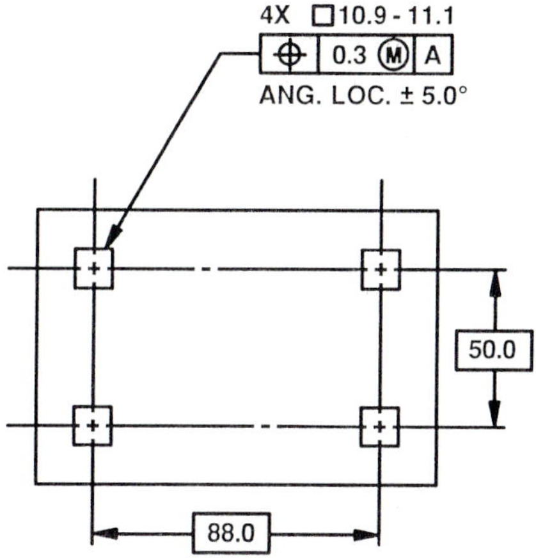

Figure 10-14. Tolerance specification allowing limited rotation of square holes

KEYS AND KEYWAYS

Since *keys and keyways* are functional features and are usually made to rigid tolerance specifications, position tolerances can be used to advantage in describing their requirements. As in previous applications, the surface of the key or keyway is controlled by a boundary or virtual size, although, the tolerances may be interpreted as an axial tolerance zone. Keyways are subject to squareness errors in two planes, both of which are controlled by position tolerances, as shown in Figure 10-16. The depth of the keyway may also be controlled by position tolerance, although it is more commonly controlled by directly toleranced dimensions, as shown in Figure 10-17.

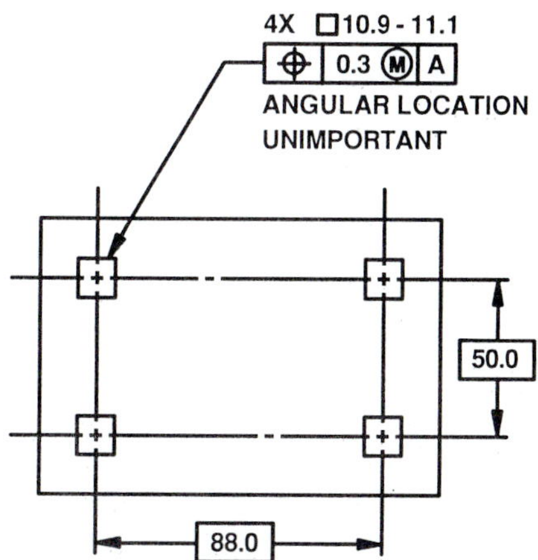

Figure 10-15. Tolerance specification allowing complete freedom of rotation of square holes

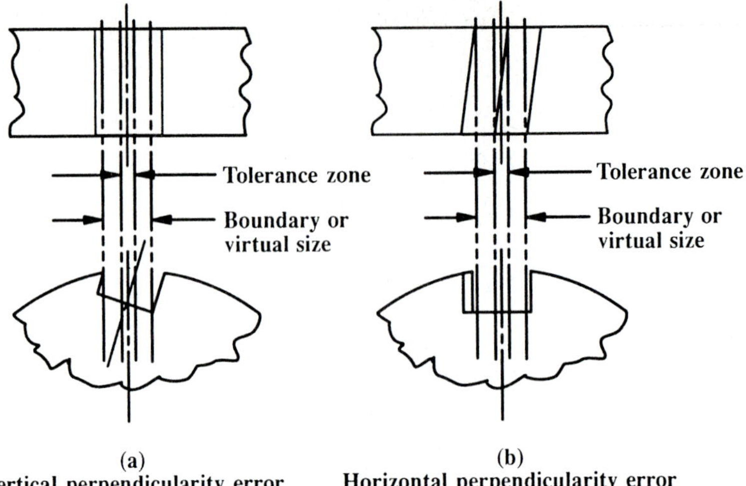

Figure 10-16. Perpendicularity errors in keyways

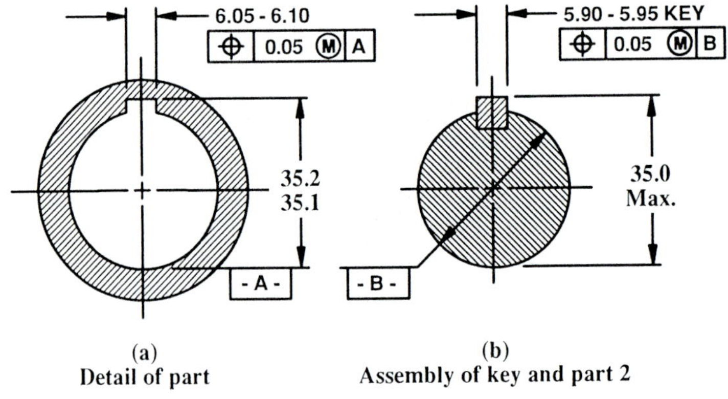

Figure 10-17. Position tolerance specification for a keyway

Example 10-16

In the figure shown, the key size is 5.95 – 6.00. If the keyway in both Parts 1 and 2 is 6.05 – 6.10, calculate balanced tolerances.

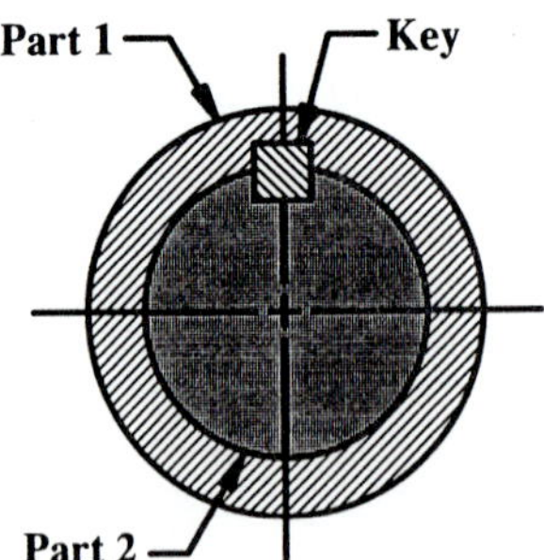

Solution Since both keyways have a nominal clearance at MMC, the key is a floating fastener and the tolerance can be derived form Equation 10-4a:

$$T_B = 6.05 - 6.00 = 0.5$$

Example 10-17

Referring to the preceding example, assume a keyway size of 6.10 – 6.15 for Part 1 and assume that the key is press fit into Part 2. If a position tolerance of 0.03 is assigned to Part 1, calculate a tolerance for the assembled key and shaft.

Solution Since the key is secured in Part 2, it constitutes a fixed fastener application and can be solved by application of Equation 3-5a.

$$T_T = 6.10 - 6.00 = 0.10$$

$$T_2 = T_T - \mathrm{T}_1 = 0.10 - 0.03 = 0.07$$

Note that the keyway in Part 2 will have a projected height tolerance zone since the key press fits into the part. Since such specifications are somewhat difficult to inspect, a practical solution would be to assign the calculated tolerance to the assembly of the shaft, Part 2, and the key.

PILOTED FASTENERS

A *piloted fastener* is defined as an application where a fastener, or the member into which it mounts, is positioned in a pilot hole and then secured (e.g., piloted weld nuts). Piloted fasteners are of two types: (1) the single element type, and (2) the multiple element type.

Perpendicularity is not a function of the hole placement for such fasteners, but rather of the welding operation. However, the fastener is assumed to remain square when calculating component position tolerances. An assembly specification can be used to check perpendicularity once the fasteners have been installed. Because gaging of pilot holes is difficult, an inspection of the assembly may be the most practical method to control both perpendicularity and position errors of pilot holes. Note that the function of the pilot hole is to restrict the location of the fastener; hence, the pilot holes are critical at their LMC size, and the boundary lies outside of the hole. This application, except for the added movement of the fastener in the pilot hole, is essentially a fixed fastener application.

Single Element Piloted Fasteners

When a fastener has clearance and pilot surfaces which are common, as in Figure 10-18a, it is called a *single element piloted fastener*.

The assembly of such a fastener can be assured by placing the LMC pilot hole so that is within the MMC clearance hole, as shown in Figure 10-18b. This prevents the fastener from interfering with the clearance hole regardless of its position with respect to the pilot hole. Then:

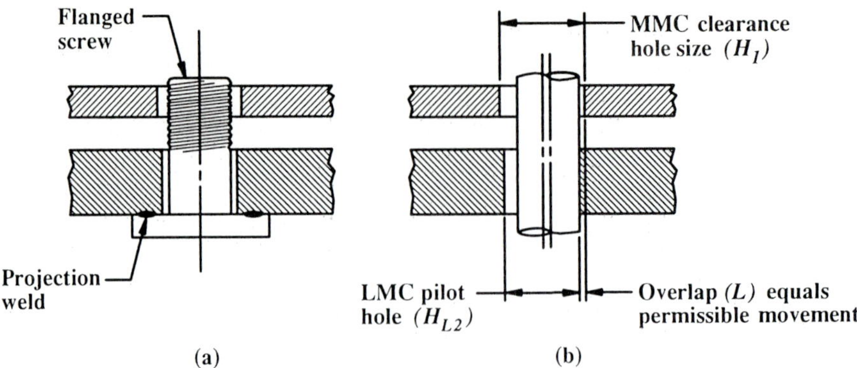

Figure 10-18. Single element piloted fastener application

$$T_T = 2L = H_1 - H_{L2} \tag{10-13}$$

$$T_B = L = \frac{H_1 - H_{L2}}{2} \tag{10-14}$$

where: H_{L2} = LMC pilot hole size
L = Overlap of clearance hole relative to pilot hole

One of the interesting characteristics of this fastener application is that the size of the fastener is important only in that it must be small enough to enter the holes.

Multiple Element Piloted Fasteners

This classification represents the general case and includes any piloted fastener which cannot be classified as a single element type. A typical example, shown in Figure 10-19, is the piloted projection weld nut and a bolt assembled into two sheet metal parts.

This fastener is subject to the influence of many variables, including:

1. Shift of the pilot in relation to the true center of the pilot hole. Since the function of the pilot is to restrict movement, not provide clearance, the critical size will be at LMC, as this allows maximum movement of the pilot.
2. Eccentricity of the fastener body to the pilot diameter. This error will displace the body with respect to the true center of the pilot hole.
3. Mislocation of the pilot hole.
4. Mislocation of the clearance hole.

By determining an assembled eccentricity, E_T, between the centerline of the fastener body and the centerline of the pilot hole (see Part 2 of Figure 10-19), the equations for fixed fasteners with eccentricity may be used.

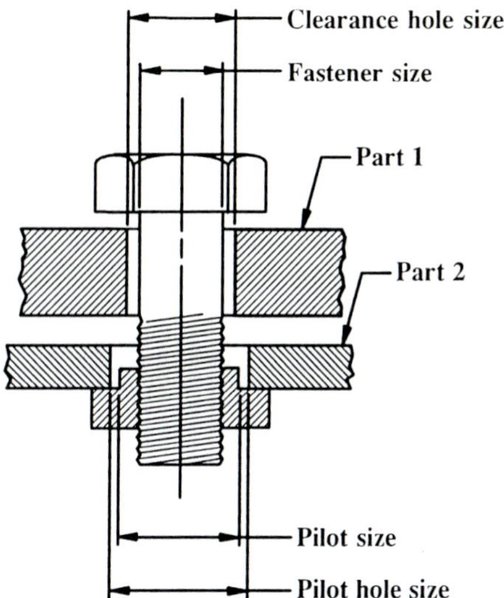

Figure 10-19. Multiple element piloted fastener application

Analyzing Figure 10-20, we see that the assembled eccentricity$_T$, equals the thread to pilot eccentricity, E, plus one-half of the diametral clearance of the pilot at LMC, Q.

$$E_T = E + Q$$

But: $C = 2E$, or $C_T = 2E_T$

$$C_T = 2(E + Q) = C + 2Q$$

But: $Q = \dfrac{H_{L2} - F_L}{2}$

$$C_T = C + (H_{L2} - F_{L2})$$

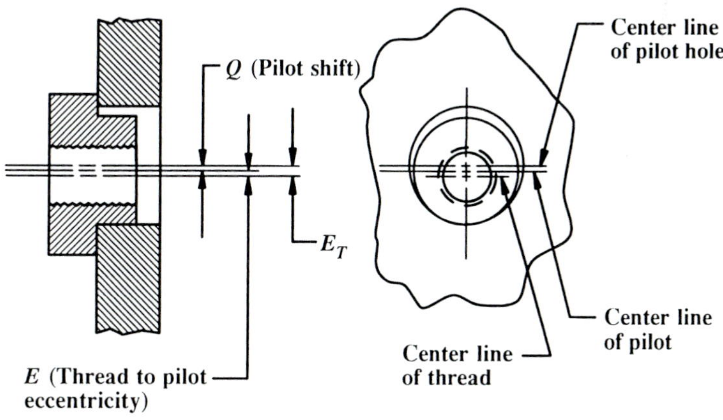

Figure 10-20. Analysis of assembled eccentricity for a multiple element piloted fastener

Substituting into Equation 10-5b, where C_T replaces C:

$$T_T = H_1 - (F_1 + C) - (H_{L2} - F_{L2})$$

$$T_T = (H_1 - F_1) - (H_{L2} - F_{L2}) - C \qquad (10\text{-}15)$$

Substituting the same data into Equation 10-6b yields the relation for balanced tolerances:

$$T_B = \frac{(H_1 - F_{1)} - (H_{L2} - F_{L2}) - C}{2} \qquad (10\text{-}16)$$

where: H_1 = MMC clearance hole size
F_1 = MMC fastener size
H_{L2} = LMC pilot hole size
F_{L2} = LMC pilot size
C = Concentricity of pilot to fastener

Assembly Requirements

Referring again to Figure 10-20, we note that a fastener may move by an amount, T_F, which is equal to the assembled eccentricity, E_T, plus the position tolerance of the pilot hole, T_C

$$T_F = T_C + 2E_T$$

$$\text{But}, E_T = \frac{C + (H_{L2} - F_{L2})}{2}$$

Then by substitution:

$$T_F = T_C + (H_{L2} - F_{L2}) + C \qquad (10\text{-}17)$$

Whenever an assembly tolerance, T_F, is specified, a projected height must be given.

Example 10-18

Given a 6mm bolt mounting into a weld nut and the following conditions:

Pilot diameter = 8.90 – 9.10

Pilot hole diameter = 9.10 – 9.40, position tolerance = 0.30

Clearance hole = 7.80 – 8.30

Thread to pilot concentricity = 0.4TIR

Solve for position tolerance at the pilot hole.

Solution From Equation 10-15:

$$T_T = (H_1 - F_1) - (H_{L2} - F_{L2}) - C$$
$$T_T = (7.80 - 6.0) - (9.40 - 8.90) - 0.4 = 0.9$$
$$T_2 = T_T - T_1 = 0.9 - 0.3 = 0.6$$

Example 10-19

Using the data of Example 10-18 and a concentricity tolerance of 0.2TIR from bolt thread to bolt body, solve for balanced tolerances.

Solution C = Sum of all concentricities = 0.4 + 0.2 = 0.6

From Equation 10-12;

$$T_B = \frac{(H_1 - F_1) - (H_{L2} - F_{L2}) - C}{2}$$

$$T_B = \frac{(7.80 - 6.00) - (9.40 - 8.90) - 0.6}{2} = 0.35$$

Example 10-20

The following data applies to Figure 10-18a: Clearance hole diameter = 8.0 – 8.5, and pilot hole diameter = 6.0 – 6.5. Solve for balance tolerances.

Solution From Equation 10-14:

$$T_B = \frac{8.0 - 6.5}{2} = 0.75$$

Example 10-21

From the data of Example 10-18 derive an assembly specification. The thickness of the part containing the clearance holes = 20mm.

Solution From Equation 10-17:

$$T_F = T_c + (H_{L2} - F_{L2}) + C$$
$$T_F = 0.3 + (9.40 - 8.90) + 0.4 = 1.2$$

The resulting assembly specification is:

⌖	⌀ 1.2 Ⓜ	DATUMS
20.0 Ⓟ		

CAGED NUTS

Caged nuts are nuts which are secured in a sheet metal panel to make the holding of the nut unnecessary during assembly. Although a number of different designs are in use, most of those available mount in a square or rectangular hole and have some clearance or *float.* There are two ways to tolerance the holes for such nuts. When it is desired that the tapped hole on the nut be completely exposed, so as to facilitate the starting of its bolt, LMC critical holes are specified and the formulas for multiple element piloted fasteners can be applied. Since both square and round tolerance zones will be involved, stationary fastener methods or correction factors will be needed. If bolt starting is not a problem, and the cage nut can be aligned by use of dog point fasteners, or other means, MMC critical holes and floating fastener principles can be applied. This has the advantage of easier gaging of the MMC holes. In either case, perpendicularity is not a problem because errors are absorbed in the nut *float*.

CONE NUTS AND BOLTS

Many varieties of *cone nuts and bolts* may be encountered; however, it is characteristic of all such applications that the fastener is secured by a thread or press fit and the conical surface of the bolt or nut is self-aligning. Figure 10-21 illustrates two types of wheel nuts which typify this application.

Clearance holes in such parts can be tolerances by conventional techniques. Since the bolt and nut have no clearance surfaces when the cone is seated, there can be no position tolerance; however, tolerances established on the basis of experience can be used for these components. Obviously, tolerance values should be relatively small because all tolerances must be absorbed by deformations in the assembled parts.

TRUSS HEAD SCREWS

In the appliance and automotive industries, there are many applications in which *truss head screws* or *oval head screws and finishing washers* are used to cover large clearance holes in sheet metal and trim panels in mechanisms such as latches and certain types of trim. Since

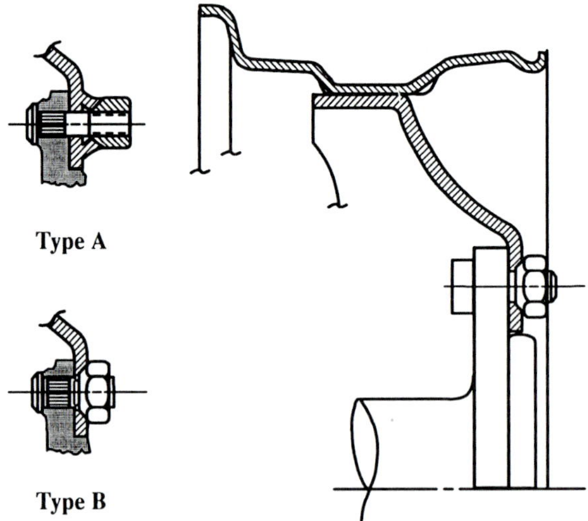

Figure 10-21. Automotive wheel nuts. Assembly typifies cone nut application

the purpose of the large clearance holes in such components is to permit their adjustment or alignment when they are assembled, only a portion of the clearance is used to determine the position tolerance applied to the holes. A simple cumulation study of the fastener size and the hole size tolerances quickly reveals whether the screw or washer is sufficiently large to cover the oversized hole in the component being secured.

From Figure 10-22, a simple formula can be derived by equating the LMC head size to the parameters which describe the body of the screw and the clearance hole.

$$H_L = \frac{F_L + HD_L}{2} - L \tag{10-18}$$

If it is assumed that sufficient overlap is allowed so that the body to head eccentricity can be ignored, the resulting Equation is:

$$H_L = \frac{F_L + HD_L}{2} - L \tag{10-19}$$

In terms of the head size required for a stated hole size:

$$HD_L = 2(H_L + L) - F_L \tag{10-20}$$

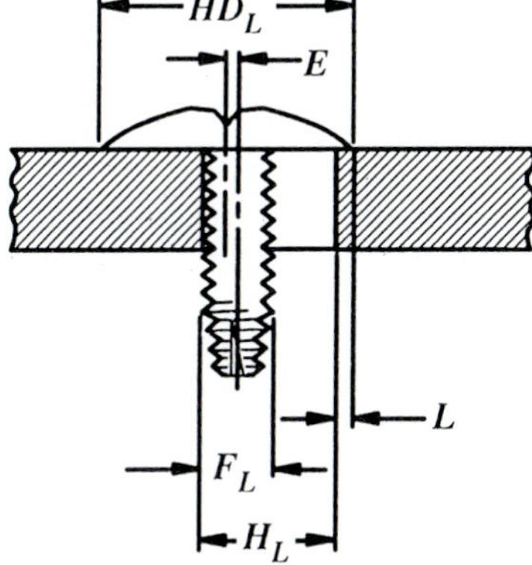

Figure 10-22. Truss head screws covering large clearance holes

EXTENSION OF FEATURE SIZE TOLERANCE

In conventional position tolerance specifications, it is a common practice to assign some finite value for position tolerance when the feature is at its critical size. It may be reasonably argued that the critical size is arbitrary and that the feature size should be limited only by the boundary if the feature is favorably positioned. To the contrary, it can be argued that feature size requirements are more easily maintained than are locations, and that expanding the size limits will have little value because a corresponding restriction in position tolerance is required. In either case, it is a question of judgment whether or not extending the size tolerance will be of value. An increasing number of authorities feel that the feature size limits should be extended in the drawing specifications so that greater latitude is allowed to the manufacturing engineer in interpreting drawings before beginning actual production.

MMC Holes

It is possible for an MMC clearance hole to be produced either undersize or oversize. An oversize hole decreases the bearing under the head of the fastener, and since there is no convenient limit to the amount of a hole may be oversized, it is not practical to generalize about this type error. Oversize holes may prove acceptable, but each occurrence must be examined and analyzed individually. On the other hand, there is a definite limit to the amount the undersized hole may be decreased. This limit is the boundary or virtual size, which is the size required to permit the fastener, as positioned by the mating part, to assemble. Figure 8-3 repeated here for convenience, illustrates the virtual size limit.

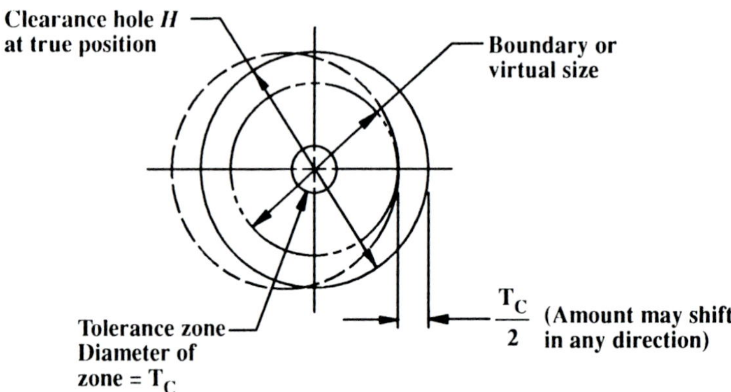

Figure 8-3. Cross section through clearance hole. Moving hole in contact with boundary causes center to generate tolerance zone

The type of application also affects the amount a hole may be undersized. If a hole is positioned exactly at its true position, as in Figure 8-3, then it is possible for the hole size to decrease until it is the same size as the boundary. By setting the hole size equal to the virtual size in Equation 8-1, we confirm that the position tolerance decreases to zero.

$$V_H = H - T_C \tag{8-1}$$

But $H = V_H$ in this example, therefore:

$$H = H - T_C\text{, hence } T_C = 0$$

If the hole size limits are expanded to include the boundary or virtual size, the position tolerance specification is written so that the stated lower limit of the hole size is the boundary size and the position tolerance is given a zero value. As an example:

A clearance hole for any fixed fastener may be reduced to it boundary size, as calculated by Equation 8-1. This is not true, however, of all floating fastener applications. In cases where the tolerances are unbalanced, or where on clearance hole is smaller than another, the boundary or virtual size may be smaller than the size of the fastener which must assemble through the hole. In these cases, the hole size may decrease only to the MMC fastener size[1] and the position tolerance is reduced by a corresponding amount. Combining Equations 9-1 and 9-2 and equating the actual hole to the fastener size yields the related position tolerance.

$$T_A = T_C + T_{ADD}\text{, or} \qquad (9\text{-}1)$$
$$T_{ADD} = T_A - T_C$$
$$T_{ADD} = H_A - H \qquad (9\text{-}2)$$
$$\text{Let } H_A = F\text{, Then}$$
$$T_{ADD} = F - H$$

Equating T_{ADD} from Equation 9-1 to T_{ADD} for Equation 9-2:

$$T_A - T_C = F - H\text{, or}$$
$$T_A = T_C + F - H \qquad (10\text{-}21)$$

For such applications, the position tolerances associated with the smallest acceptable hole size is used for the tolerance specification.

Example 10-22

Calculate the position tolerance requirement for Part 1 of Example 10-1 using extended size tolerance. For this example: $T_C = 0.7$, $H = 10.5$, and $F = 10.0$.

Solution From Equation 10-21

$$T_A = T_C + F - H$$
$$T_A = 0.7 + 10.0 – 10.5 = 0.2$$

RFS Holes

Since hole size is a functional consideration in all RFS applications, no generalization regarding allowable error in hole size can be made. Extended size tolerance is not appropriate for this category.

LMC Holes

Undersize holes are not usually acceptable for LMC applications because the clearance for piloted fasteners usually approaches zero at the minimum hole size. Undersize holes may be acceptable in some parts, but each part must be analyzed individually , and no generalization can be made. Oversize holes, equal or smaller than the virtual size are always acceptable. As with fixed fastener applications, unbalancing the position tolerances does not

[1]For applications where eccentricity is involved, use the virtual fastener size.

affect this relation, and zero limit tolerances can be specified for all LMC applications. The maximum allowable hole size is the boundary or virtual size, which can be determined from Figure 12-5 for a given tolerance specification.

$$V_H = H_L + T_C \qquad (10\text{-}22)$$

MMC Projections

As previously noted, MMC projections are functionally similar to LMC holes and, therefore, a zero limit position tolerance can be established for all projection applications. The maximum allowable projection size is the virtual size, which may be calculated by use of Equation 8-2.

SUMMARY

The basic categories of position tolerance application include *floating, fixed,* and *piloted fasteners*. These basic concepts can be expanded to allow for fastener eccentricity, or positive clearance at assembly, through the use of *virtual fastener size*. The concept of virtual fastener size leads to the conclusion that fasteners with stepped bodies should be controlled by MMC position tolerance specifications.

Specific categories of position tolerance application include *counterbores, carriage bolts, keys* and *keyways, caged nuts, cone nuts and bolts,* and *truss head screws*.

Analysis of position tolerance applications indicates that feature size may be extended when position tolerances are reduced. This is accomplished by reducing the position tolerance to zero, and making the critical feature size equal to the boundary or virtual size.

Chapter 11

Position Tolerance Conversions

Although functionally derived position tolerances for round holes are diametral, many drawing tolerances, inspection techniques, and machine controls are based upon coordinate measurements. Therefore, it is important to understand the relationship between measured rectangular coordinates and diametral measure and to be able to make the necessary conversions.[1] This relationship between the two systems of measurement is shown in Figure 11-1.

COORDINATE TOLERANCE CONVERSIONS

If a hole location is measured and found to deviate from its true position dimensions (*A* and *B*) by some values (*X* and *Y*, respectively as shown in Figure11-1a) the diametral measurement may be calculated from Figure 11-1b and Equation 11-1. Note that the deviation *X* and *Y* may lie on either side of the true position lines.

$$D = 2\sqrt{X^2 + Y^2} \qquad (11\text{-}1)$$

Values for *D* may be charted as in Table 11-1. By use of a computer, a table of this nature can easily be generated for any range of sizes and for any chosen increments.

Another method used to convert rectangular coordinate into diametral measurement is given in Figure 11-2. To determine the diametral measurement (*D*) with this chart, find

[1]Most CMM computers contain a subroutine which can convert measured coordinate deviations to diametral (polar) values.

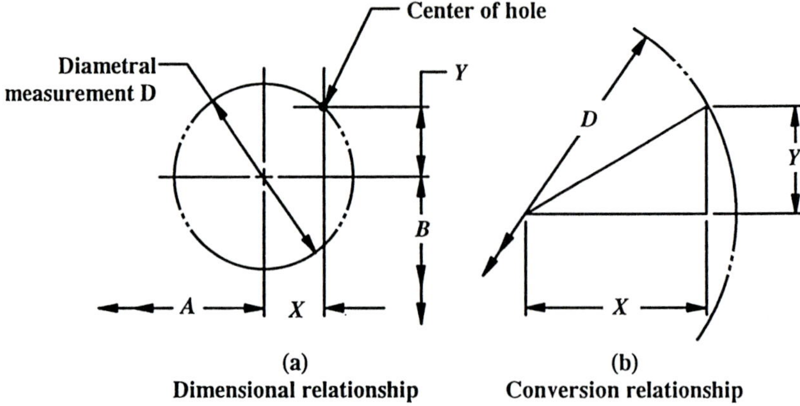

Figure 11-1. Conversion of true position deviations from rectangular coordinates to diametral measurement

the intersection of the vertical and horizontal gridlines corresponding to the deviations X and Y, respectively. Read the value of D from the circular line passing through this intersection. For example, assume the deviations of a hole location are: $X = 0.6$; $Y = 0.8$. From Figure 11-2, the circular line passing through these coordinates yields $D = 2.0$.

QUADRILATERAL TOLERANCE CONVERSIONS

It may be necessary to convert paired quadrilateral drawing tolerances to equivalent diametral position tolerances when tolerancing new parts to mate with existing components. A machinist often needs equivalent quadrilateral tolerances for a specified diametral position tolerance. Quadrilateral tolerance deviations, X and Y as shown in Figure 11-1b, are equal. The formulated relation between coordinate and diametral measure becomes:

$$D = 2\sqrt{2}X = 2.828X, \text{ and} \tag{11-2}$$

$$X = \frac{D}{2\sqrt{2}} = .353D \tag{11-3}$$

Table 11-1. Diametral tolerance values

Vertical Tolerance Deviation (Y)					
0.5	1.02	1.08	1.17	1.28	1.41
0.4	0.82	0.89	1.00	1.13	1.28
0.3	0.63	0.72	0.85	1.00	1.17
0.2	0.45	0.57	0.72	0.89	1.08
0.1	0.28	0.45	0.63	0.82	1.02
0.0	0.1	0.2	0.3	0.4	0.5

Horizontal Tolerance Deviation (X)

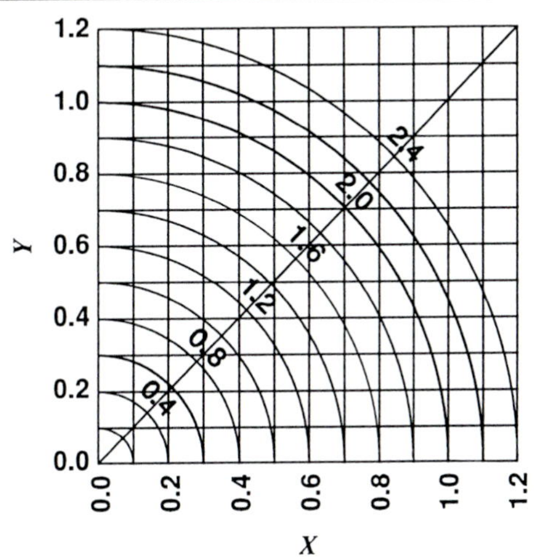

Figure 11-2. Rectangular coordinate conversion to diametral measure

Values for D may be read from the "staircase" section of Table 11-1. Values of X corresponding to even values of D are easily calculated and charted in any size range and increment through the use of Equation 11-3.

Figure 11-2 may also be used for this conversion. If a quadrilateral tolerance is given, follow either grid line corresponding to its value to the diagonal line. Read the value of D from the circular line passing through this intersection. If a position tolerance is given, follow the circular line corresponding to its value to the diagonal line and read the equivalent quadrilateral tolerance from the intersecting grid line. For example, a quadrilateral tolerance of ± 0.5 corresponds to a diametral tolerance of ∅ 1.4.

Since quadrilateral tolerances are not an accurate functional description of round hole location requirements, caution must be used when converting them to position tolerances. When making such conversions, as in the case of updating drawings based upon quadrilateral tolerances, examination of actual parts, fixtures, and gages may be desirable. When hole location gages for an existing component are available, they are the best source of information for determining the position tolerance actually used. Because this type of gage is based upon position tolerance principles, inspection of the actual gage, or examination of the gage drawing, will allow the exact position tolerance used to be determined by applying the principles outlined in Chapter 15 on fixtures and gages.

CONVERSION TECHNIQUES

For patterns of only two holes, any dimension between holes has a quadrilateral tolerance of twice the quadrilateral tolerance for each hole. This results from summing the tolerance for the two holes. If the holes are on a common centerline, the quadrilateral tolerance for

Quadrilateral Tolerance to Position Tolerance	Hole Pattern	Position Tolerance to Quadrilateral Tolerance
T = B Example: Given B = 1.0; T = 1.0	A - 2 holes 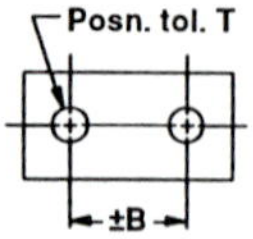 Posn. tol. T ±B	B = T Example: Given T = 0.5; B = 0.5
T = 1.414 B Or use value of X to find T in Fig. 11-2 Example: Given B = 1.2; X = 0.6 and T = 1.7 from Fig. 11-2	B - 2 holes 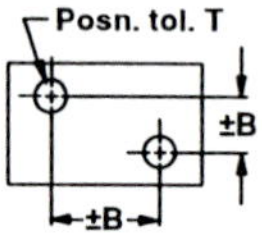 Posn. tol. T ±B ±B	B = .707 B Or use value of T to find X in Fig. 11-2 Example: Given T = 1.6; X = 0.56 from Fig. 11-2 B = 2(0.56) = 1.12
T = 2.828 B Or use value of B to find T in Fig. 11-2 Example: Given B = 0.5; T = 1.4 from Fig. 11-2	C-3 holes 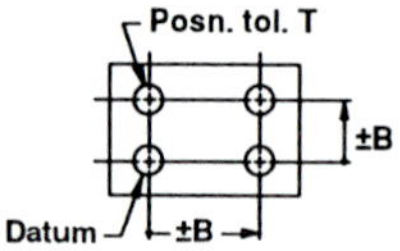 Posn. tol. T ±B Datum ±B	B = .354 T Or use value of T to find B in Fig. 11-2 Example: Given T = 1.2; B = .42 from Fig. 11-2

Figure 11-3. Chart of conversion techniques for various hole patterns

each hole is equal to one-half the position tolerance, as deviation must occur along the common centerline.

For patterns of three or more holes, the quadrilateral tolerance for each hole becomes the tolerance of its location dimension. Such tolerances cannot be summed due to accumulation of errors. Figure 11-3 summarizes quadrilateral conversions of position tolerances.

SUMMARY

Conversion techniques are used to convert position tolerances for round holes from diametral to coordinate, or coordinate to diametral measure. They are used to permit evaluation of inspection measurements, machining on coordinate systems, and the updating of old drawing specifications. Figures 11-2 and 11-3 provide ready aids for making such conversions.

Chapter 12

Cumulative Effect of Position Tolerance

In addition to maintaining the requirements of interchangeable assembly, control of the amount of material between adjacent holes, or between the edge of a part and a hole or projection, may be necessary. For example, the manufacturing requirements of a punching or drilling operation may be such that a minimum amount of material must be maintained between holes. Strength requirements may also dictate control of the minimum amount of material surrounding a hole.

While position tolerances do not directly exercise this type of control, their use permits the calculation and subsequent analysis of such relations. Using this technique, a stress analysis of a critical section could be based upon the minimum amount of material between holes. A stack-up of tolerances affecting the location of the surface might consider both the minimum and maximum amount of material between a feature and the edge of a part. Other calculations of this nature are performed with ease.

Two fundamental concepts are used to analyze all tolerance cumulation problems. These concepts are *virtual* and *resultant size*. Virtual and resultant sizes are concentric limits describing the points closest to, and furthest from, the nominal position that a surface of a feature may occupy. The virtual size concept, introduced in Chapter 8, will be supplemented in this chapter by the resultant size concept which is developed as the limit of positions opposite the virtual size.

Round features are used to illustrate the principles of tolerance cumulation, principally as a convenience, but also because most applications of tolerance cumulation theory will be to round features. The equations developed are equally applicable to other feature shapes which are symmetrical, such as square holes, elongated slots, and the like. The principles of tolerance cumulation can be applied to irregular shapes, but before attempting to apply any equation developed here to such features, its development should be reviewed to be sure of its applicability.

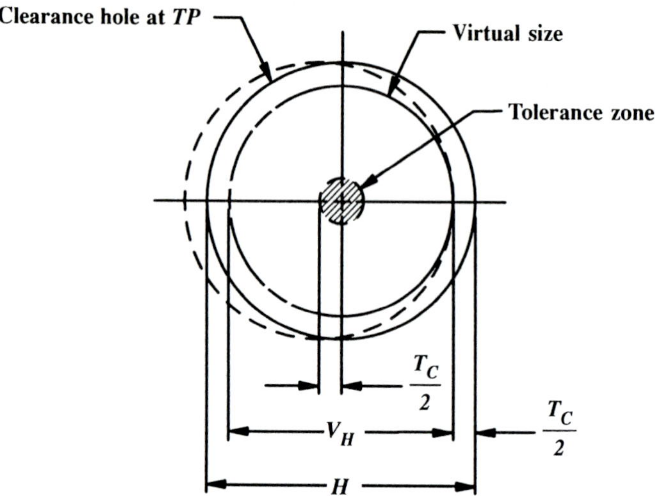

Figure 12-1. Virtual size for MMC critical round holes

MMC APPLICATIONS

As shown in Figure 8-1, the virtual size for MMC applications is associated with the division of assembly clearance. Hence, virtual size for a clearance hole is equal to, or smaller than, the critical hole size, while virtual size for an assembled fastener or projection is equal to or larger than the projecting feature. Figures 12-1 and 12-2 illustrate MMC virtual size for holes and projections, respectively. These sizes are given by Equations 8-1 and 8-2 reproduced from Chapter 8.

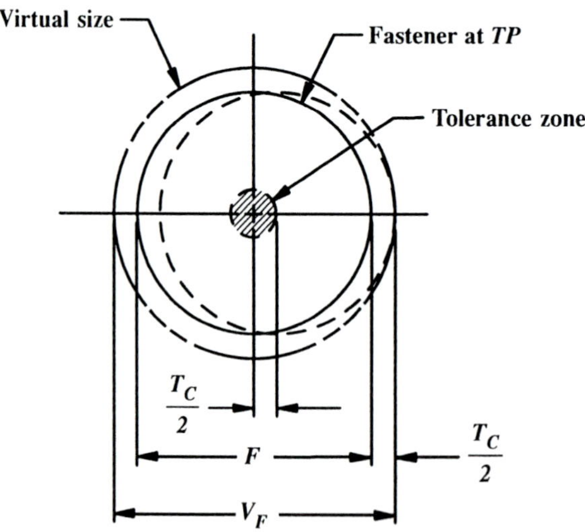

Figure 12-2. Virtual size for MMC critical round fasteners and projections

$$V_H = H - T_C \tag{8-1}$$

$$V_F = F + T_C \tag{8-2}$$

Since the virtual size for clearance holes is constructed within the hole, the resultant size is constructed in the material surrounding the hole. For this reason, the resultant size for a clearance hole is larger than the hole, while the resultant size for a fastener or projection is smaller that the projecting feature. Figure 12-3 illustrates how the resultant size for a clearance hole is developed by shifting the largest sized hole possible, H_L, against the virtual size, V_H, and then circumscribing the resultant Size, R_H.

$$\frac{R_H}{2} = H_L - \frac{V_H}{2}$$

$$\begin{aligned} & R_H &= 2H_L - V_H \\ \text{but} \quad & V_H &= H - T_C \text{ by using Equation 8-1} \\ & R_H &= 2H_L - H + T_C \\ & R_H &= 2H_L + T_C - H \qquad (12\text{-}1) \end{aligned}$$

When a position tolerance specification is given for a fastener assembled into a mounting holeor for a specific projection, the resultant size is constructed by moving the smallest size projecting feature, F_L, against the virtual size, V_F, and then inscribing the resultant size, R_F, as in Figure 12-4.

$$\frac{R_F}{2} = F_L - \frac{V_F}{2}$$

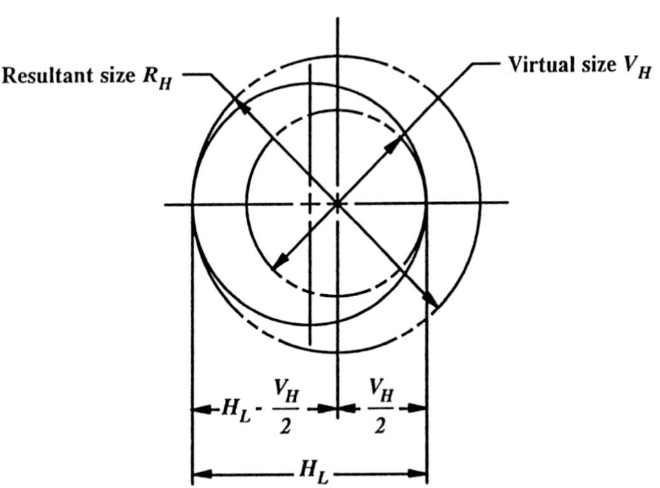

Figure 12-3. Resultant size for MMC critical round holes

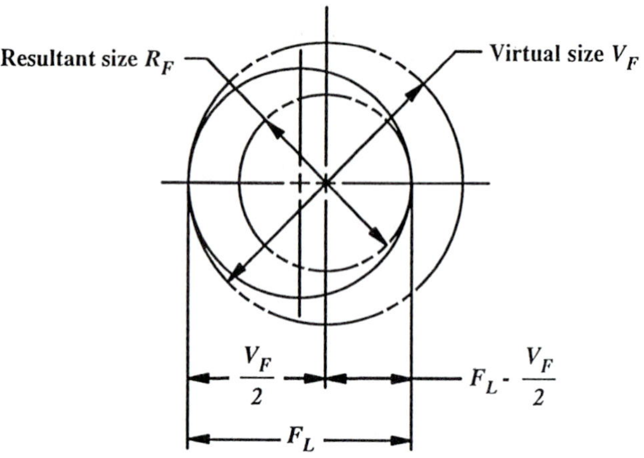

Figure 12-4. Resultant size for MMC critical round projecting features

$$R_F = 2F_L - V_F \text{, but } V_F = F + T_C \text{ by using Equation 8-2}$$
$$R_F = 2F_L - F - T_C$$
$$R_F = 2F_L - T_C - F \qquad (12\text{-}2)$$

LMC APPLICATIONS

For LMC holes, the virtual and resultant sizes are developed in a manner similar to that used for MMC holes. Since the maximum deviation of the surface of an LMC hole is the critical factor for determining if mating parts can be interchangeably assembled, the virtual size is outside the hole. The virtual size, given by Equation 10-22, is illustrated in Figure 12-5. Note the similarity to Figure 12-2 and Equation 8-2.

$$V_H = H_L + T_C \qquad (10\text{-}22)$$

The resultant size, illustrated by Figure 12-6, is developed by moving the smallest hole size, H, against the virtual size, V_H, and inscribing the resultant size, R_H. Again note the similarity of the resultant size, given by Equation 12-3, to the resultant size for projecting features, given by Equation 12-2.

$$R_H = 2H - T_C - H_L \qquad (12\text{-}3)$$

RFS AND MOUNTING HOLE APPLICATIONS

For RFS and other mounting holes,[1] the virtual size is not uniform over its full length as are other virtual size applications, but is equal to the boundary size only at the interface surface of the hole. Hence, its maximum virtual size will be the same as described for MMC fasten-

[1] As discussed in Chapter 9, RFS holes and tapped holes may be considered as a single category, designated "mounting holes."

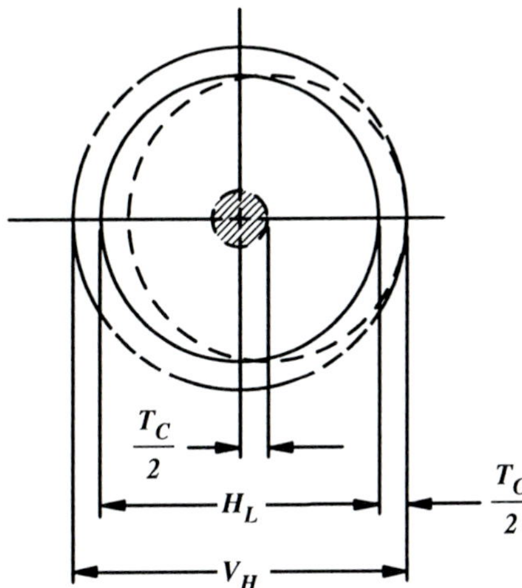

Figure 12-5. Virtual size for LMC critical

ers plus an amount due to possible perpendicularity errors which extrapolate beyond the projected boundary (or tolerance zone) into the mounting hole. As shown in Figure 12-7, the maximum virtual size may be derived by substituting H_L for F in Equation 8-2 and adding the increase in virtual size ($2d$) due to extrapolated perpendicularity error.

$$V_H = V_H \text{ (from Equation 8-2)} + 2d$$
$$V_H = H_L + T_C + 2d$$

where: $d = \frac{T_C D}{P}$, then

$$V_H = H_L + T_C + \frac{2T_C D}{\mathrm{P}}$$

$$V_H = H_L + T_C \left[1 + \frac{2D}{P}\right] \tag{12-4}$$

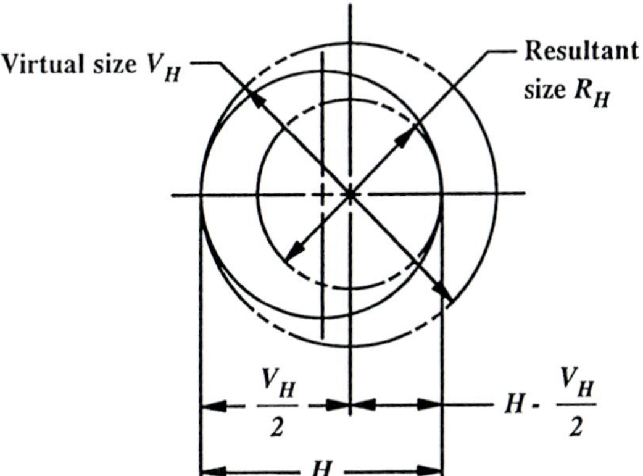

Figure 12-6. Resultant size for LMC critical holes

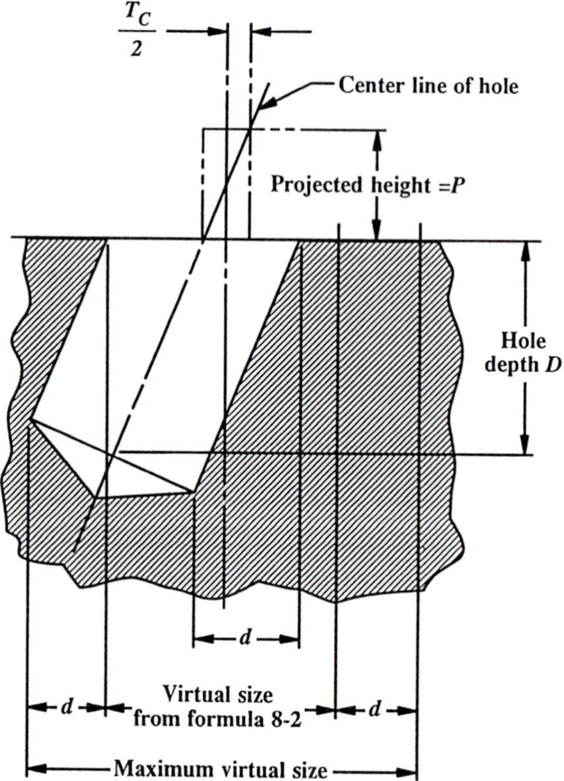

Figure 12-7. Virtual size for mounting holes

The minimum resultant size for mounting holes is the size inscribed within the smallest hole size when it has been displaced by the full amount of its position error. Except for the effect of extrapolated perpendicularity, this resultant size is similar to the virtual size for MMC holes, given by Equation 8-1.

$$R_H = H - T_C$$

Correcting for perpendicularity extrapolation shown in Figure 12-7:

$$R_H = H - T_C - \frac{2T_C D}{P}, \text{ or}$$

$$R_H = H - T_C \left[1 + \frac{2D}{P}\right] \qquad (12\text{-}5)$$

To summarize all the relations developed, Table 12-1 lists the inner and outer limits describing the position of symmetrical features.

Table 12-1. Inner and outer limits of a positionally toleranced feature.

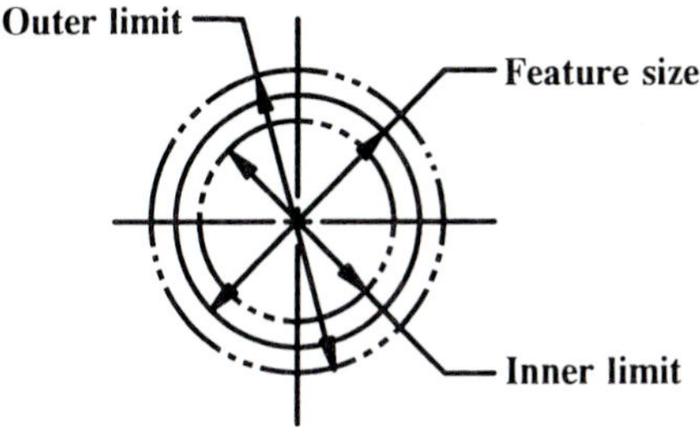

Feature	*Critical Size*	*Outer Limit*	*Outer Limit*
Holes	MMC	$2H_L + T_C - H$	$H - T_C$
	LMC	$H_L + T_C$	$2H - T_C - H_L$
	RFS and tapped	$H_L + T_C\left(1 + \frac{2D}{P}\right)$	$H - T_C\left(1 + \frac{2D}{P}\right)$
Projection	MMC	$F + T_C$	$2F_L - T_C - F$

SOLVING TOLERANCE CUMULATION PROBLEMS

The following examples illustrate how the relations specified in Table 12-1 may be used to solve tolerance cumulation problems.

Example 12-1

Find the minimum and maximum stock between holes indicated by Dimension *A*.

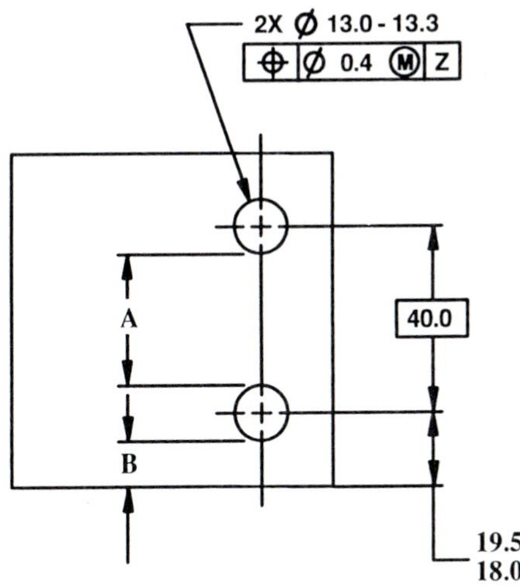

Solution A_{min} = Dimension between holes – Outer limit of holes. Noting that the stated critical size is MMC.

$$
\begin{aligned}
\text{Outer limit} &= 2H_L + T_C - H \\
&= 2(13.3) + 0.4 - 13.0 \\
&= 26.6 + 0.4 - 13.0 = 14.0 \\
A_{min} &= 40.0 - 14.0 = 26.0
\end{aligned}
$$

A_{max} = Dimension between holes – Inner limit of holes.

$$
\begin{aligned}
\text{Inner limit} &= H - T_C \\
&= 13.0 - 0.4 = 12.6 \\
A_{max} &= 40.0 - 12.6 = 27.4
\end{aligned}
$$

Example 12-2

Resolve Example 12-1 for Dimension A if the holes are $M12 \times 1.5 - 6H$ threads in a 25.0 thick part with a projected height of 20.0

Solution

$$
\begin{aligned}
\text{Thread major diameter} &= 12.0 \\
\text{Outer limit} &= H_L + T_C \ [1 + \frac{2D}{P}] \\
&= 12.0 + 0.4\,[1 + \frac{2 \times 25.0}{20.0}] \\
&= 12.0 + 0.4(3.5) = 13.4 \\
\text{Inner limit} &= H - T_C\,[1 + \frac{2D}{P}] \\
&= 12.0 - 0.4(3.5) = 10.6
\end{aligned}
$$

Using the same equation as in Example 12-1:

$$
\begin{aligned}
A_{min} &= 40.0 - 13.4 = 26.6 \\
A_{max} &= 40.0 - 10.6 = 29.4
\end{aligned}
$$

Example 12-3

Referring to the figure of Example 12-1, solve for the minimum and maximum edge to hole distance, Dimension B.

Solution B_{min} = Minimum distance of edge to hole-one-half of outer limit of hole.
From Example 12-1:

$$
\begin{aligned}
\text{Outer limit} &= 14.0 \\
\text{Inner limit} &= 12.6\text{, then} \\
B_{min} &= 18.0 - \frac{14.0}{2} = 11.0
\end{aligned}
$$

B_{max} = Maximum distance of edge to hole – one-half of inner limit of hole.

$$
B_{max} = 19.5 - \frac{12.6}{2} = 13.2
$$

Example 12-4

Calculate the minimum and maximum values for Dimension C.

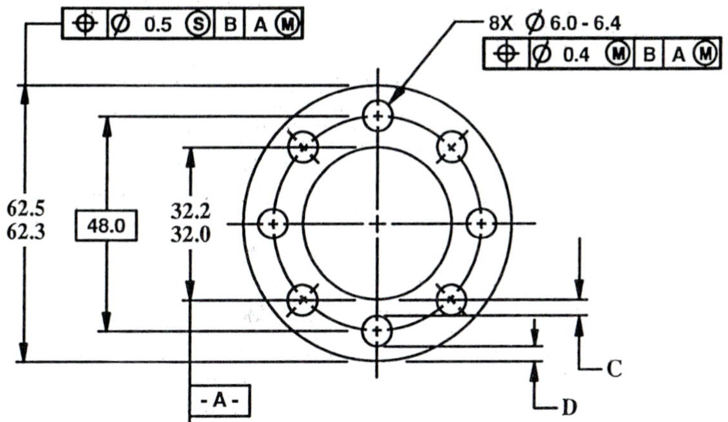

Solution The datum, Diameter A is an MMC hole with a zero position tolerance value, then

$$C_{min} = 1/2(\text{Bolt circle diameter} - \text{Outer limit of holes} - \text{Outer limit of pilot})$$
$$C_{max} = 1/2(\text{Bolt circle diameter} - \text{Inner limit of holes} - \text{Inner limit of pilot})$$

For the holes:

$$\text{Outer limit} = 2H_L + T_C - H = 2 \times 6.4 + 0.4 - 6.0 = 7.2$$
$$\text{Inner limit} = H - T_C = 6.0 - 0.4 = 5.6$$

For the pilot:

$$\text{Outer limit} = 2 \times 32.2 + 0 - 32.0 = 32.4$$
$$\text{Inner limit} = 32.0 - 0 = 32.0$$

Then:

$$C_{min} = 1/2(48.0 - 7.2 - 32.4) = \frac{8.4}{2} = 4.2$$

$$C_{max} = 1/2(48.0 - 5.6 - 32.0) = \frac{10.4}{2} = 5.2$$

Example 12-5

Recalculate Example 12-4 if the position tolerance specification for the holes reads:

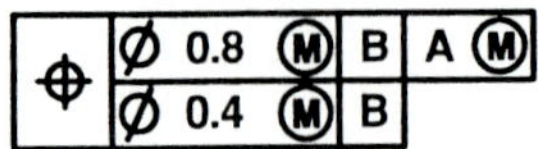

Solution In the case of a composite position tolerance, the pattern location portion represents the worst case condition between the toleranced features and their datums, in this case the 6.0 – 6.4 holes and the 32.0 – 32.2 pilot diameter. Hence, the new limits for the holes becomes:

$$\begin{aligned}\text{Outer limit} &= 2 \times 6.4 + 0.8 - 6.0 = 7.6 \\ \text{Inner limit} &= 6.0 - 0.8 = 5.2\end{aligned}$$

Continuing as in Example 12-4:

$$\begin{aligned}C_{min} &= 1/2(48.0 - 7.6 - 32.4) = \frac{8.0}{2} = 4.0 \\ C_{max} &= 1/2(48.0 - 5.2 - 32.0) = \frac{10.8}{2} = 5.4\end{aligned}$$

Example 12-6

Calculate D_{min} and D_{max} as shown in the illustration of Example 12-4.

Solution

D_{min} = 1/2(Minimum OD – Bolt circle diameter – Outer limit of hole) – Shift of bolt circle relative to OD.

D_{max} = 1/2(Maximum OD – Bolt circle diameter – Inner limit of hole) + Shift of bolt circle relative to OD.

Data for the holes from Example 12-4:

$$\begin{aligned}\text{Outer limit} &= 7.2 \\ \text{Inner limit} &= 5.6\end{aligned}$$

The shift of the bolt circle relative to the OD = Eccentricity of bolt circle to Diameter A + Eccentricity of eccentricity of OD to Diameter A. Solving for maximum datum location tolerance by Equations 9-3 and 9-4:

$$\begin{aligned}T_A &= T_C + T_{ADD} \\ T_{ADD} &= H_L - H = 32.2 - 32.0 = 0.2 \\ T_A &= 0 + 0.2 = 0.2 \text{ for bolt circle} \\ T_A &= 0.5 + 0.2 = 0.7 \text{ for OD} \\ \text{Shift} &= 1/2(0.2 + 0.7) = 0.45\end{aligned}$$

Then:

$$\begin{aligned}D_{min} &= 1/2(62.3 - 48.0 - 7.2) - 0.45 = 3.1 \\ D_{max} &= 1/2(62.5 - 48.0 - 5.6) + 0.45 = 4.9\end{aligned}$$

Example 12-7

Two pieces are assembled with bolts which project above the finished assembly to serve as mounting studs. Using data given for the components, determine an assembly specification that could be used to develop the tolerances for a bracket that will mate with the studs (projecting bolts) of the finished assembly.

Data: The bolt is 12mm in diameter. The two mating parts use balanced tolerances of Ø 0.5, hole size of 12.5 – 13.5, and an MMC specification.

Solution The bolts can be positioned against the outer limits of the holes:

$$\text{Outer limit} = 2(13.5) + 0.5 - 12.5 = 15.0$$

Am MMC bolt placed against the outer limit would have a position tolerance equal to the amount above the fastener size:

$$F + T_C = 15.0$$
$$T_C = 15.0 - 12.0 = 3.0$$

Example 12-8

Repeat the calculation of Example 12-5 on a probability basis.

Solution: An obvious choice would be to conduct a Monte Carlo simulation. Lacking that capability, an approximation can be made as follows:

From the example: $C_{min} = 4.0$

$C_{max} = 5.4.$ or $C = 4.7 \pm 0.7$

Four tolerances contributed to this stack-up: hole size, hole location, pilot size, and hole pattern to pilot location.

From Chapter 3, assuming approximately equal tolerance values:

$T_T = \sqrt{\Sigma t_i^2} = \sqrt{4t^2} = 2t$ on a probable basis.

$T_T = 4t$ on a linear basis.

Hence, probability 50% of linear value.

then; $C = 4.7 \pm (0.7 \times .5) = 4.7 \pm 0.35$

A Benderizing factor could be applied if a bias to the nominal locations was expected.

SUMMARY

Analysis of the cumulative effect of position tolerances permits the calculation of the variable distance between features, between a feature and the edge of a part, or other factors that relative to size or strength of a part. Although formulated as individual cases, Table 12-1 summarizes all possible cases to facilitate problem solving.

Chapter 13

Virtual Size

Sheet metal parts, weldments, and other fabricated structures often require alignment of holes in two or more parts or sections. Most tolerance standards do not discuss such applications. Some of the unique features to these applications are:

1. Such parts may contain a single hole rather than a pattern of holes.
2. Specification of holes by nominal size and size tolerance is not important because holes are already sized in the component parts.
3. When there is a pattern of holes, location is assumed accurate in the components and only needs to be main tained in the assembly operation. That is, the main concern is with location of the pattern rather than individual holes.

The approach outlined here uses conventional position tolerance for component parts and virtual size requirements for assemblies. The virtual size notation represents the size of a true geometric shape that can be inscribed in a hole assembled from two or more parts (see Figure 13-1). The specifications are such that they are easily calculated and translated in fixture, gage, or inspection requirements. For example, the virtual size may be used as a *go* pin size for a gage or fixture. This concept may be applied to any shape hole.

Note that hole alignment specifications may be used in lieu of or as a supplement to hole location dimensions. Inclusion of such specifications, however, does not exclude the use of dimensions necessary to define the location of a hole. virtual size specifications are best used when holes are critical at MMC (i.e., holes used as clearance holes). Under such conditions, virtual size readily describes assembly conditions.

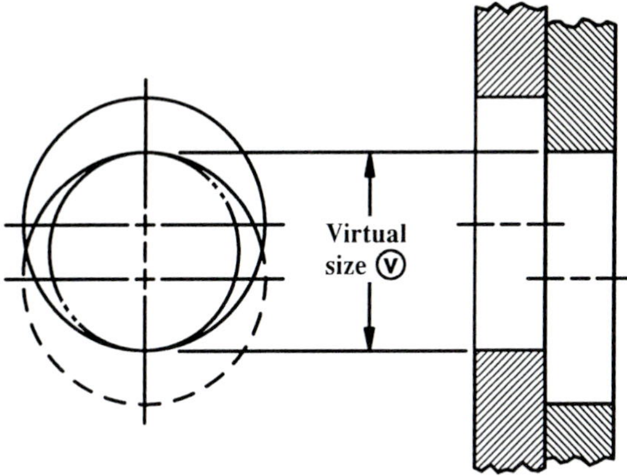

Figure 13-1. Virtual size inscribed in a hole assembled from two parts

Figure 13-2 shows three hole alignment requirements. In Figure 132a, a hole alignment requirement is used instead of locating dimensions. In Figure 13-2b, the lugs welded to the plate typify the application of attitude tolerance to hole alignment requirements. In this case, the large location tolerance is restricted so that the paired holes cannot be displaced in opposite directions. In Figure 13-2c, the clevis is an example of a part where hole alignment is not required since the holes are drilled in a single operation. The specification note is selected to emphasize that the two holes are considered to be a single surface, although interrupted.

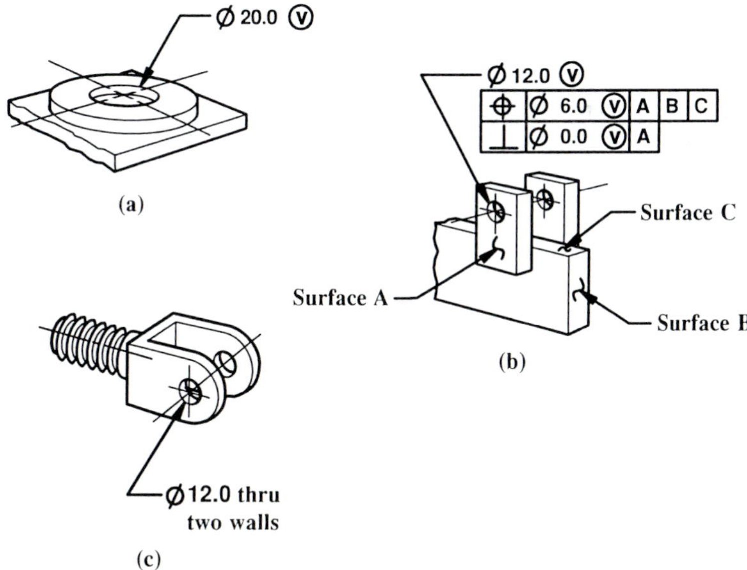

Figure 13-2. Three typical applications. These applications demonstrate variation in hole alignment requirements

The basic specification (see Figure 13-2a) controls only the alignment of the holes involved. Other, more complex specifications (see Figure 13-2b) control alignment as defined by a datum. In all cases, product function is the basis for determining the type of specification to be used.

THE BASIC ALIGNMENT PROBLEM

Many assemblies require only the entry of a certain size component in the unit. This situation is typified by doubler plates into which a bolt, hitch pin, or other fastener must be inserted. The outstanding characteristic of such a structure is that alignment is required only for the pin-type component. Such assemblies often are used to connect chain, cables, or adjustment devices to a framework. Frequently, they contain only one hole. The virtual size is readily determined to be the same as the MMC size of the smallest of the mating parts. If a minimum clearance must be maintained, the virtual size must be checked against the MMC size of the mating part to ensure the desired clearance. A typical application and specification is shown in Figure 13-2a.

ALIGNMENT TO A DATUM

Many assemblies require not only that a fastener enter component parts, but that it placement be restricted to some nominal position. This typically occurs when assembled parts are fitted together with small clearances. The outstanding characteristic of such assemblies is their inability to absorb errors due to misalignment of holes. As with a single hole assembly, virtual size is determined from the MMC sizes of the component parts.

The weldment of Figure 13-2b, shown in its assembled form in Figure 13-3, requires a specification which applies attitude controls to achieve the necessary hole alignment. In this case, the chosen control is perpendicularity, however, parallelism is also appropriate in some cases. Figure 13-5 shows that a specification without the perpendicularity requirement might permit the entry of Pin *A*, but would not allow Arm *B* to assemble.

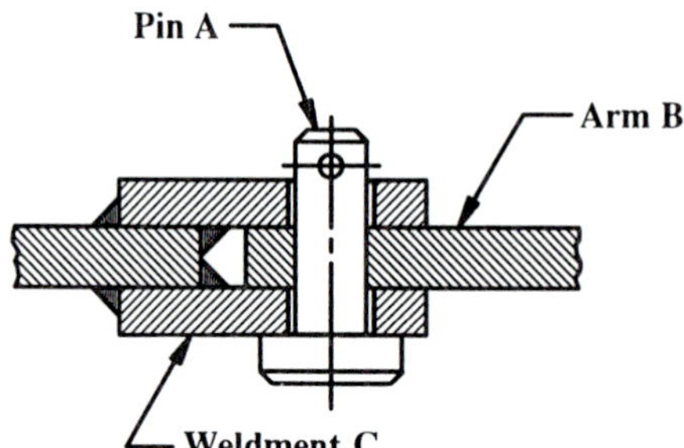

Figure 13-3. Clevis type assembly requiring both hole alignment and attitude control

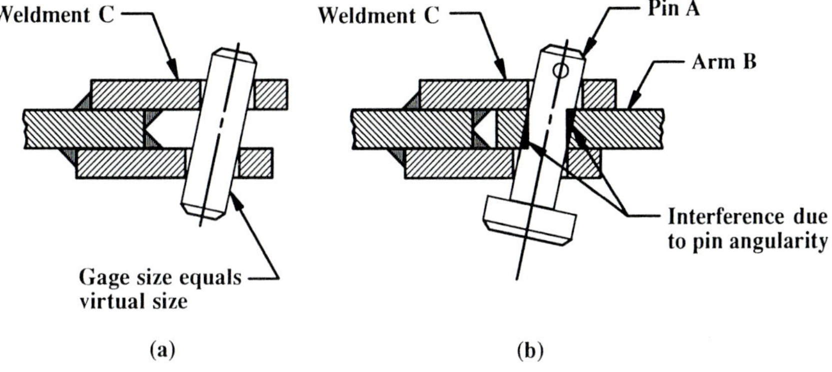

Figure 13-4. Hole alignment specification without perpendicularity. This specification permits pin to enter (a) but leads to interference on assembly (b)

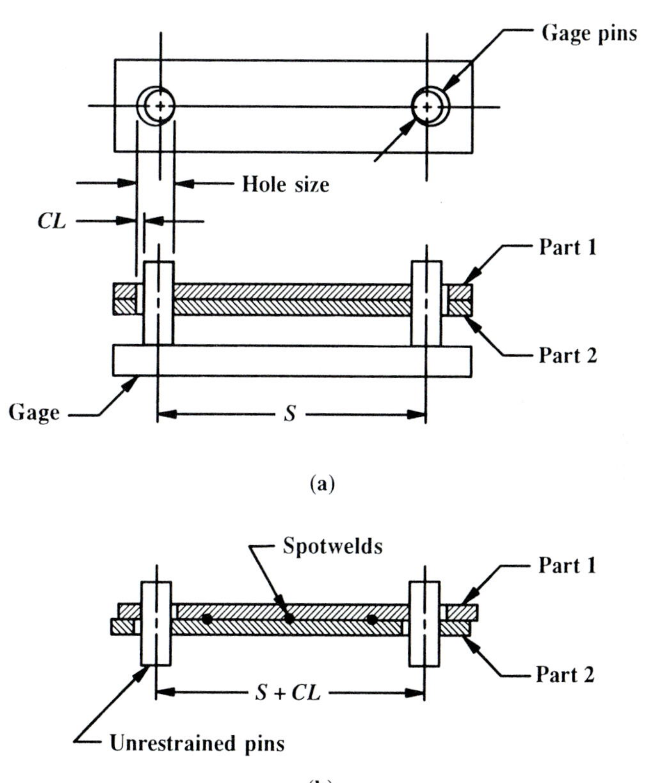

Figure13-5. Center distances controlled to ensure alignment. Proper fixturing (a) can prevent assembly errors caused by unrestrained dimensioning (b)

ALIGNMENT WITHIN A PATTERN

When parts containing holes which constitute a single hole pattern are joined, dimensional location of the holes, in addition to alignment, must be maintained to ensure assembly with mating parts. Here control is necessary to avoid accumulation of detail errors. Figure 13-5 shows two parts with maximum spacing between holes. These parts are checked by a composite gage with pins sized to a predetermined virtual size diameter and spaced at the nominal center distance S. If the center distance S is not restrained, it could be possible for either part 1 or part 2 to shift by the amount of the clearance CL. The center distance would then be expanded to $S + CL$. Many other combinations of detail and assembly error are possible if center distances are not maintained.

To ensure accurate alignment of any pattern of holes, it is necessary to align, or fixture, each hole at its appropriate location. If parts are produced from the same tools, RFS alignment at two (widely spaced) holes should be sufficient.

For patterns of holes, the virtual size must be based upon the MMC hole size and the critical position tolerance of the detail parts. Reexamination of Chapter 8 will show the virtual size is given by Equation 8-1:

$$V_H = H - T_C \tag{8-1}$$

A typical specification based upon use of these principles is shown in Figure 13-6.

PROCEDURE FOR ANALYZING ASSEMBLIES

To simplify the analysis of parts containing hole patterns, the following procedure may be used:

1. Tolerance component parts with conventional position tolerance techniques and specifications. Disregard subsequent operations, such as welding, at this point.

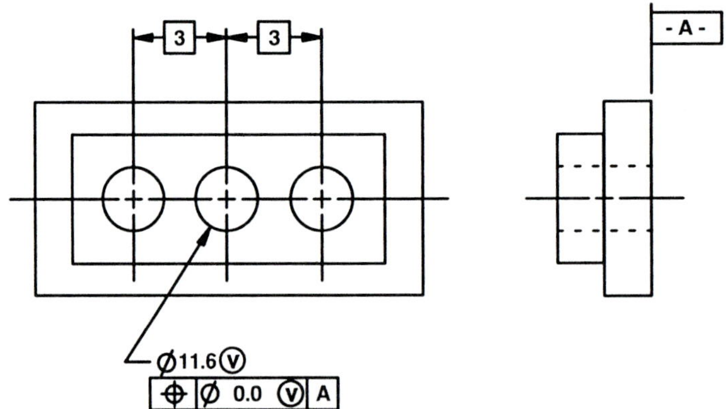

Figure 13-6. Alignment of each hole in an assembly

2. Calculate virtual sizes from Equation 8-1. If not identical for the various components, the smallest value must be used.
3. On the assembly drawing, specify the basic hole location dimensions, the virtual size, and a zero position tolerance value at virtual size.
4. For floating fastener applications, make sure that the stationary fastener method was used to tolerance the components. Otherwise, a virtual size smaller than the component being assembled may result.

Example 13-1

The two parts shown in Figure 13-6 are to be welded together. The weldment assembles to a plate containing tapped holes for 10mm bolts. If a hole size of 10.6–11.0 is used for both parts, calculate the position tolerance for all parts and the virtual size. Make the position tolerance for the tapped holes twice that of the clearance holes.

Solution

$$\begin{aligned} \text{Let } T_1 &= \text{Tolerance for tapped holes, and} \\ T_2 &= \text{Tolerance for clearance holes.} \\ T_T &= T_1 + T_2 = 2T_2 + T_2 = 3T_2 \end{aligned}$$

From Equation 10-5a:

$$\begin{aligned} T_T &= H - F \\ 3T_2 &= 10.6 - 10.0 \\ T_2 &= \frac{0.6}{3} = 0.2 \\ T_1 &= 2 \times 0.2 = 0.4 \end{aligned}$$

From Equation 13-1:

$$\begin{aligned} V_H &= H - T_C \\ &= 10.6 - 0.2 = 10.4 \end{aligned}$$

Example 13-2: Auto Door Hinge Tolerances

In this example, the operation sequence is to fixture the hinges to the door from the outer surface and the adjacent hemmed edge and to bolt them in place. This requires that the tapped block, which is welded in place at a prior subassembly stage, have sufficiently accurate hole locations. The hinge and other parts through which the fasteners pass must also be sized and located properly. The following study determines the various sizes and tolerances.

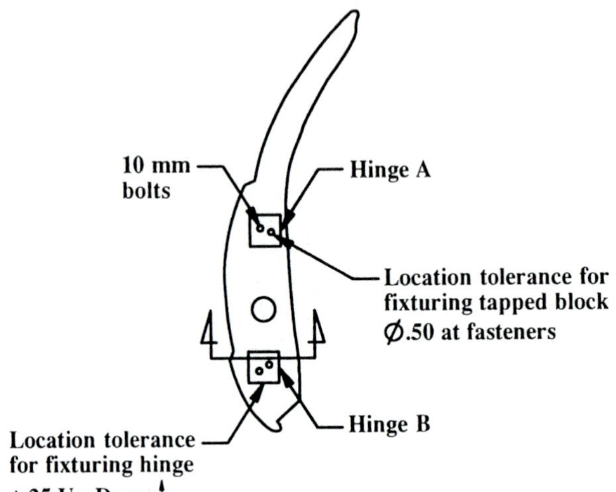
10 mm
bolts
Hinge A
Location tolerance for
fixturing tapped block
Ø.50 at fasteners
Hinge B
Location tolerance
for fixturing hinge
±.25 Up-Down
±.25 In-Out

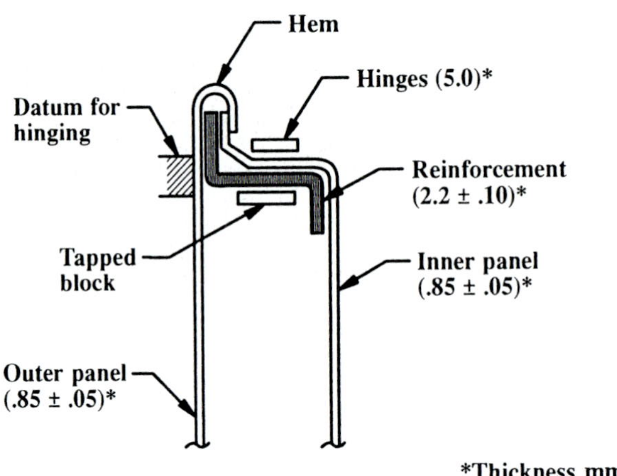
Hem
Hinges (5.0)*
Datum for
hinging
Reinforcement
(2.2 ± .10)*
Tapped
block
Inner panel
(.85 ± .05)*
Outer panel
(.85 ± .05)*

*Thickness mm

The key to doing this study in a simple and effective manner lies in another example of the use of the virtual size concept.

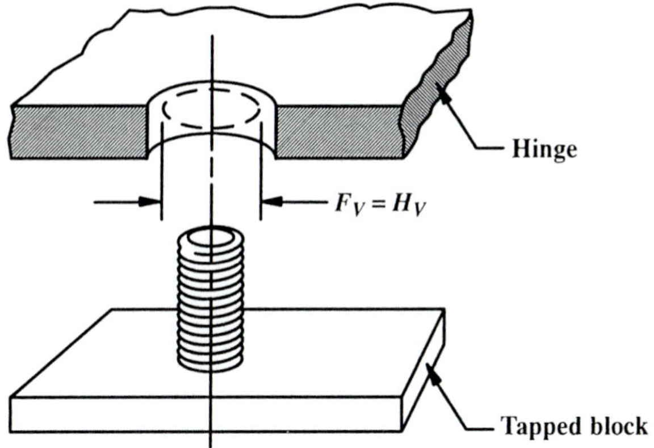

Let F_V = MMC bolt diameter + Hole location tolerance in block + Block location tolerance.

$F_V = 10.0 + T_B + 0.50 = 10.5 + T_B$

For the hinge:

H_V = MMC hole diameter – Hole location tolerance in the hinge – Hinge location tolerance.

Note that a balanced tolerance is used for the hinge and block reflecting the similar nature of the parts.

$$H_V = H - T_B - 2 \times 0.25\sqrt{2} = H - T_B - 0.70.$$

but: $H_V = F_V$, or

$$10.5 + T_B = H - T_B - 0.70$$

$$H = 2T_B + 11.2$$

Setting $T_B = 0.5$, yields

$H = 12.2$ at MMC and $F_V = 11.0$

Tapped Block Specification

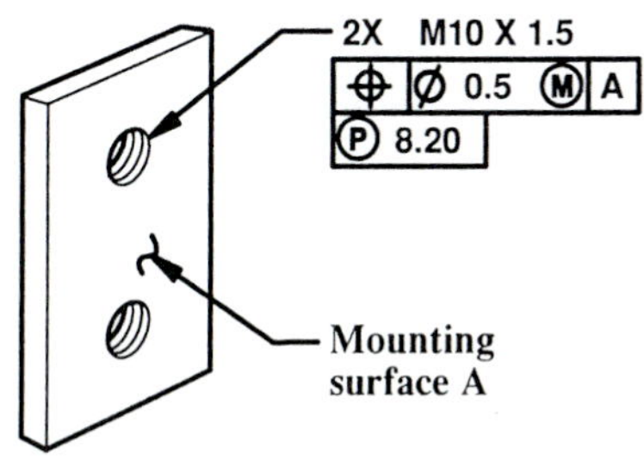

Example 13-2. Tapped block

The projected height is calculated as follows:

5.0	Hinge
2.3	Reinforcement
0.9	Inner panel
8.2	

Hinge Specification

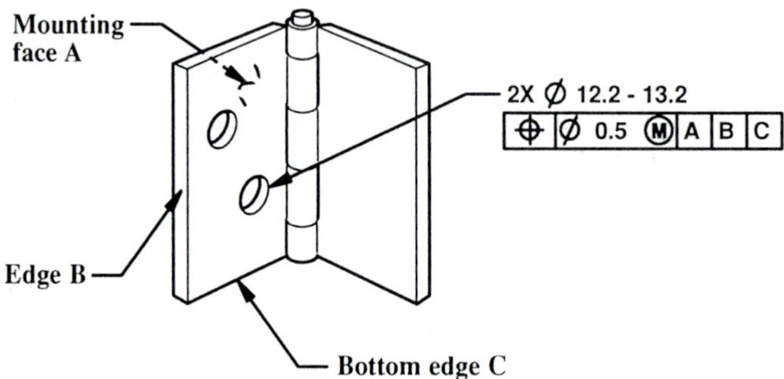

Example 13-2. Hinge

Door Assembly Specification

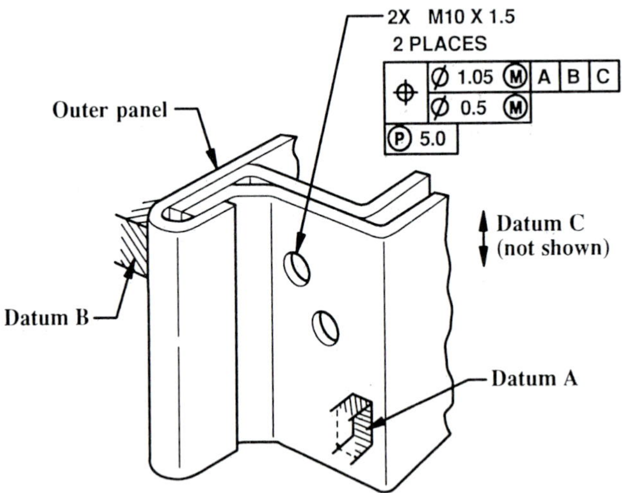

Example 13-2. Door assembly specification

The door assembly position tolerance has two elements in a composite specification. The ∅ 0.5 value is from the tapped block and becomes the tolerance *within the pattern*. The pattern location tolerance to the datums comes from:

0.50 Within pattern tolerance
0.50 fixture tolerance in 11.0 F_V value
0.05 Outer panel thickness tolerance
1.05

The projected height is reduced to the hinge thickness of 5.0 as its reference (Datum A) has shifted.

Inner Door Assembly Specification

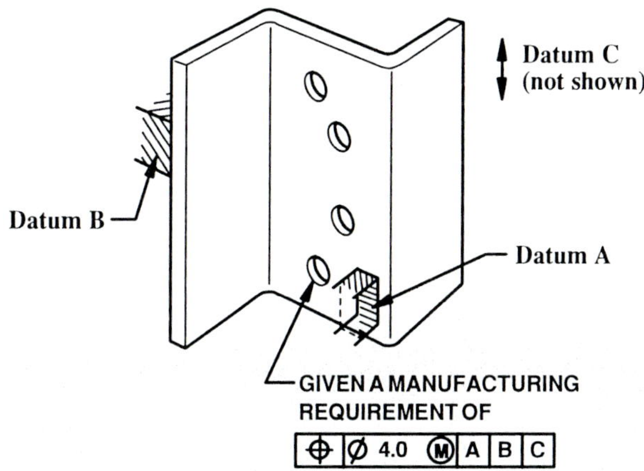

Example 13-2.

The inner door specification is identical to the door assembly specification except that the outer panel is not yet in place and its thickness variation is not a factor. Hence, the upper position tolerance in the composite specification has a value of 1.0.

Solving for hole size the total tolerance formula is used as the virtual size already includes the tolerances which affect the fastener:

$$T_T = H - F_V$$
$$4.0 = H - 11.0, \text{ or } H = 15.0 \text{ at MMC}$$

Inner Panel

The inner panel is identical to the reinforcement except that it is effected by the thickness range of the reinforcement, hence:

$$T_T = 4.0 + 0.2 = 4.2, \text{ and}$$
$$H = 15.2 \text{ at MMC.}$$

SUMMARY

Conventional position tolerances combined with virtual size notation solves complex problems not discussed in most tolerance standards. For MMC holes, control of virtual size will ensure that functional requirements are maintained. Datum and dimensional requirements may be required to complete the specifications.

Virtual size is also shown to be a powerful tool for solving problems involving hole alignment.

Chapter 14

Consistency Tolerance

Under present merchandising practices, it is not uncommon to experience situations in which interchangeability is important in assembled modules, but not at a detail part level. Pumps, timers, motors, and many similar appliance components are serviceable as modules, but their component parts are not usually available. In most of the preceding discussions, it was assumed that accurate[1] dimensions and placement of features was necessary to allow parts to be assembled or serviced. What then are the requirements for component modules such as those mentioned above.

CONSISTENCY TOLERANCE

For such parts, consistency in manufacturing operations provides the necessary control of part dimensions and feature location so as to ensure assembly and functional requirements. This consistency can also be the key to lower unit costs since it can fulfill the product requirements while allowing maximum process latitude.

The concept of manufacturing consistency is demonstrated by a typical example (see Figure 14-1). In this piece of earth moving equipment, the item of main concern is the shovel loader lift arm. The holes at the two attaching points on the twin mounting plates must be properly aligned because of the close fits required for the hydraulic cylinder pivot pins which are heavily loaded. Note that interchangeability is important at the assembly level, but the only critical requirement for the mounting plates is that Dimension *A* be con-

[1]*Accurate* is used here in a strict technical sense, meaning exact duplication of a standard unit of measure (e.g., 100.0mm).

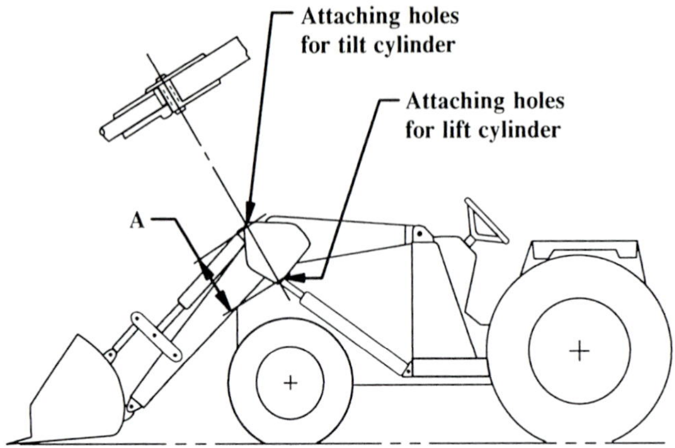

Figure 14-1. Alignment of holes in shovel loader lift arms. This alignment is critical because of required pin fits

sistent so that proper assembly alignment can be attained. A high degree of accuracy for Dimension *A* is not essential if precise alignment of the holes is maintained.

Frequently, in parts of this type, an attempt is made to hold Dimension *A* to very accurate location limits in manufacturing to achieve the necessary consistency. The plate shown in Figure 14-2a illustrates this approach to the problem. Such accuracy is often too expensive to achieve for a part of this type.

To provide a more practical way of coping with the problem, a consistency tolerance can be imposed upon a wider accuracy limit as shown in Figure 14-2b. By definition, a *consistency tolerance* requires the parts in any lot to be consistent within the specified tolerance value for the dimension so designated. In this case, Dimension *A* must be consistent within 0.04mm, however, the accuracy of the location dimension may vary by ± 1.0mm.

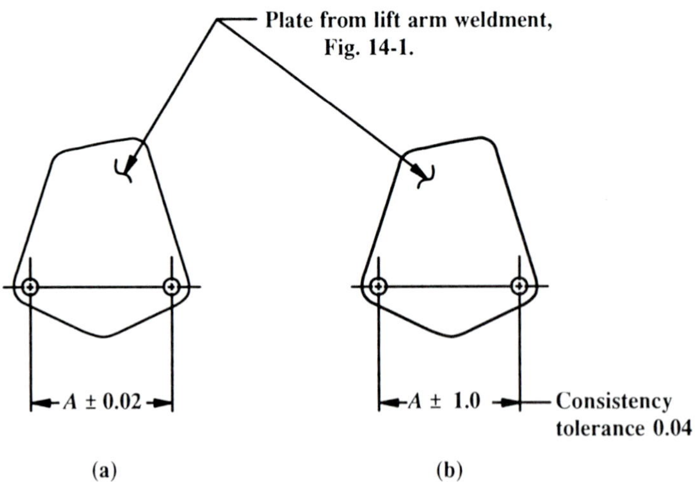

Figure 14-2. Alternate methods of tolerancing the lift arm pivot plate to obtain the required alignment

This part may now be manufactured by any process which will maintain the required consistency. For example, any one of the following processes might be used:

1. Punch both holes simultaneously in each part with a die which is used for punching all of the parts.
2. Punch holes separately in each part with a universal punching machine using a locator in one punched hole to maintain consistency in the position of the second punched hole.
3. Punch holes through two parts which are tack welded or clamped together.
4. Line drill the holes after the plates are welded to the lift arm.

Method 2 in this list serves to illustrate the control of consistency in lots. If the part is manufactured in lots, say 100 each month, the accuracy of Dimension *A* may vary between lots, if parts from separate lots are not allowed to be mixed. Therefore, a particular setup on the punching machine would not have to be made to close dimensional limits. However, if lots must be mixed, a process tolerance may be applied, tightening the accuracy to maintain the required consistency. This process requirement is represented by the practice shown in Figure 14-2a. For parts made with lot segregation, Figure14-2b applies. Inspection would entail comparison of parts selected a random from within a lot.

Although the example used here to illustrate the concept of Consistency tolerancing deals with hole location, the same reasoning can be applied to any tolerance type. Hole locations can also be a combination of position and Consistency tolerances.

SUMMARY

Consistency tolerances are used to provide a method by which parts can be made to consistent dimensions without the need for unwarranted accuracy. Their application permits the adoption of less expensive manufacturing techniques. Inspection of such parts entails the comparison of parts selected at random.

There are no national standards or symbols for consistency tolerances. Companies that elect to apply them will need to develop an internal standard.

Chapter 15

Fixtures and Gages

Fixtures and gages[1] serve related, but distinctly different purposes. *Fixtures* are devices designed and built to position either a component or the tools, such as drills, punches, or the like, which are used in the machining or forming of various features of the component. The use of *gages* is one of the many techniques by which the completed component may be verified to be acceptable according to its engineering specifications. Fixtures and gages for a given component are often similar in construction, using, for example, the same method to hold or position a component. Thus, it can be seen that many of the same principles apply to their design.

Occasionally, a certified fixture is used to perform the functions of both gage and fixture. This is accomplished by periodic inspection of the fixture to certify its accuracy, thereby negating, in some instances, the need for independent gaging operations. Alternately, a fixture may be used to hold a part during an inspection operation (e.g., as a support and datum system used with a *coordinate measuring machine*).

It will not be our purpose here to discuss the details of tool design, but rather to show the need for maintenance of continuity in design analysis, drawing specifications, and tool design.

[1]Manufacturing engineers often distinguish between *fixtures* and *jigs,* the former being a device to locate and hold a part while the latter is a device to locate and control the tools. The word "fixture" is used here to refer to both of these functions insofar as the part requirements involve the relationship between datums (location of the part) and location of features (such as the holes produced by drills).

FIXTURE DESIGN

There are two considerations to be made in translating an engineering specification into an intelligent fixture design. The first of these is the positioning of the component under consideration, and the second involves tool positioning.

Positioning the Component

Component positioning entails an evaluation of either a datum specification, or of pattern location dimensions, and the subsequent design of locators or other devices to maintain these functional requirements. Pattern location requirements, or targeting requirements, are met by providing contact at the stated point, line, or surface (see Figure 15-1). Where pattern location dimensions are taken from component edges, the entire surface may be placed in contact with the fixture locator, or the area of contact may be restricted by using any of the techniques used with datum targets. When the latter technique is selected, care must be taken to use the same points or areas on subsequent fixtures or gages to avoid the introduction of positioning error due to form errors on the part. This is often documented on a process sheet. Datum sequence must always be considered when designing a fixture. A part must be stabilized against the various datum surfaces in the sequence that they are listed.

Figure 15-2 shows the type of locators to be used for datum specifications where internal or external pilots are involved. The nominal pilot size for MMC pilots is usually toleranced so that the fixture plug or bore cannot interfere with the component datum feature. Most corporations have standards which cover the tolerancing of such fixture features.

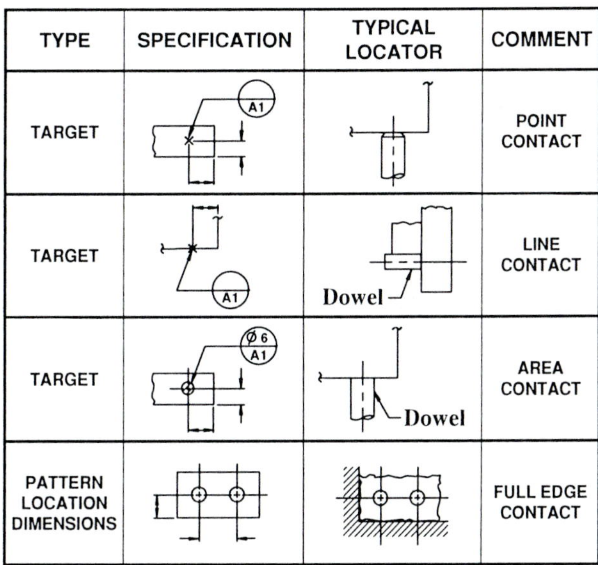

TYPE	SPECIFICATION	TYPICAL LOCATOR	COMMENT
TARGET	A1		POINT CONTACT
TARGET	A1	Dowel	LINE CONTACT
TARGET	Ø6 A1	Dowel	AREA CONTACT
PATTERN LOCATION DIMENSIONS			FULL EDGE CONTACT

Figure 15-1. Locator methods for fixture design

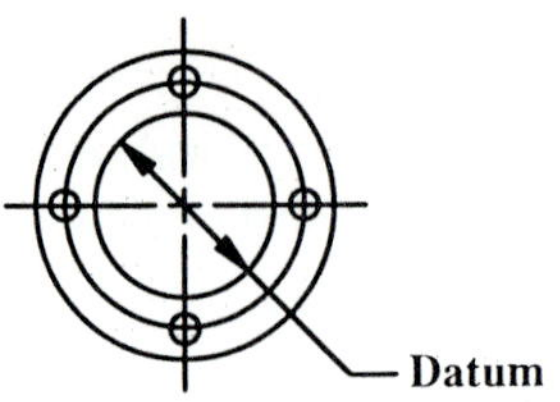

TYPE OF PILOT	TYPE OF LOCATOR	NOMINAL SIZE
INTERNAL MMC	PLUG	LOW DIMENSIONAL LIMIT OF BORE
EXTERNAL MMC	BORE	HIGH DIMENSIONAL LIMIT OF PILOT
INTERNAL OR EXTERNAL RFS	CENTRALIZING DEVICE	MUST COVER SIZE RANGE OF PILOT

Figure 15-2. Piloting methods for fixture design

In other instances, reliable texts or handbooks on tool design may be consulted. When virtual size datums are encountered, the datum critical size must be modified to include the influence of other applicable tolerances, such as position, perpendicularity, and so on.

Tool Positioning

The second consideration of most fixture designs involves the positioning of the tools used to form or machine the component features. In some cases, the feature positions are determined by moving the machine tool relative to the component and its fixture. More frequently, the feature locations are determined by the fixture through the use of drill bushings, punch guides, or the like.

In either case, the fixture design requires that the feature locations be dimensioned from the location surfaces and toleranced according to established tooling conventions. Since the machining of most fixtures is done on a *jig bore* or a similar precision machine, such locations are usually both dimensioned and toleranced by coordinate dimensions to correspond with the coordinate traverse movement of these machines. Where round tolerance zones are involved, as frequently is the case, the tolerance values are usually converted to equivalent coordinate tolerances by using the techniques of Chapter 11. Tolerance zones with special or unusual shapes require individual consideration.

Coordinate Fixture Tolerances

Once coordinate component tolerances are determined, coordinate fixture tolerances are assigned. Prior to extensive use of GD&T, a common practice was to assign a certain percentage (usually 20 percent) of the component tolerance as the fixture tolerance. However, since some tolerances, including coordinate tolerances derived from round position tolerances are variable and depend upon the size of the feature produced, rule-of-thumb assignment of tolerances for fixtures becomes more difficult.

Two alternatives are available. The first is to apply a percentage rule to the mean tolerance determined by calculating the average of the critical and maximum position tolerances. A second, more sophisticated, method is to determine the probable feature size from studies of similar processes and to determine the position tolerance which correspond to the probable feature size. A percentage rule can be applied to the resulting component tolerance. Obviously, if the component tolerances are large, the critical tolerances can be used to generate fixture tolerances without further consideration as they will generate fixture tolerances that are relatively easy to meet.

The Economic Approach

Perhaps a better approach to the problem is an economic one. Since the purpose of the fixture is to produce the desired component tolerances, all the available tolerance should be usable for manufacturing purposes. Although drill run-out, bushing clearance, and other considerations must be a part of the evaluation for available fixture tolerances, it is the author's opinion that the accurate statement of product requirements contained in GD&T specifications can permit larger tolerances to be used in fixture construction. In this vein, some companies have traditionally used 30 to 50% of the component tolerance as the fixture tolerance. Although it can be offered only as a principle, fixture tolerances should be selected to allow for other manufacturing variations, to maintain the product requirements and appropriate capability indexes, and to keep the fixture cost to a minimum. Obviously, such an approach would lead to different policies where different manufacturing techniques are involved, and perhaps even in different areas of the same corporation.

Figure 15-3 illustrates the fixture for producing the component shown in Figure 9-17.

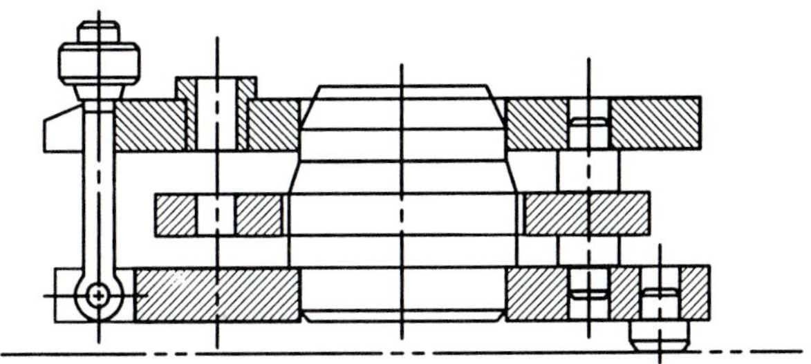

Figure 15-3. Drill fixture for component shown in Fig. 9-17

GAGE DESIGN

The primary purpose of any *gage* is to verify that the component being inspected meets engineering requirements. A second purpose may be to provide process control data.

Gage designs fall into two broad classifications . . . attribute and variable. The basic technique employed in an attribute gage is to construct the boundary requirements of the component in the form of *hard* surfaces and, by assembling the gage and the component to show that the boundary requirements of the component are not violated. With such a gage, the only information that it provides is that the part passes or fails. Size or fit is not quantified and trends usually can not be observed. *Variable gages,* on the other hand, replace the hard surfaces with *soft* ones (i.e., indicators, sensors, and other devices that allow the gaging operation to be quantified). This is the approach used where SPC is desired. Generally, the *attribute gage* is easier to use, requires less time, and is more durable, all at the expense of providing less information. If one is certain that the process output lies in a flat region of a *loss function* curve, the attribute gage is an ideal choice.

If appropriately designed, it is possible to have a gage that has indicators or sensors that interchange with hard features. This allows the gage to be used as a variable gage supporting SPC until process capability is demonstrated. At that time, the gage can be converted to attribute control for more expeditious use on the shop floor.

Position Toleranced Features

The first consideration in developing a gage is to design the portion which gages the position toleranced features. Figure 15-4 repeats the boundary or virtual size requirement for MMC features, as stated in Chapter 8. It is obvious that the gage features are similar to the features of the mating part.

It should also be noted that, where gages are available for components toleranced in

FEATURE	GAGE	SIZE
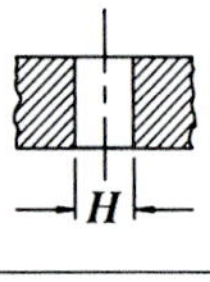 H	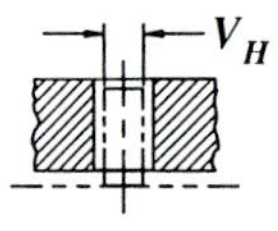 V_H	$V_H = H - T_C$
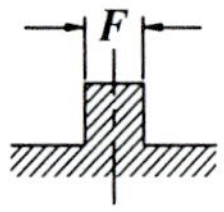 F	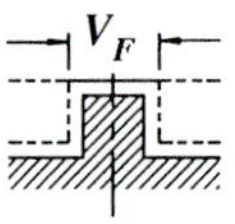 V_F	$V_F = F + T_C$
Non-uniform tolerance 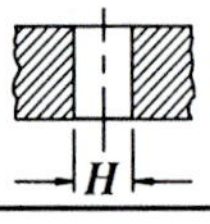 H	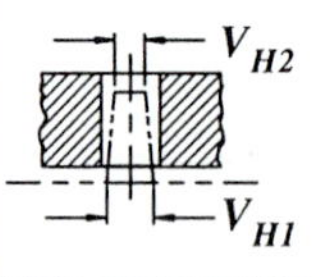 V_{H2} V_{H1}	$V_{H1} = H - T_{C1}$ $V_{H2} = H - T_{C2}$

Figure 15-4. Gage features for MMC components

the old coordinate tolerance system, the position tolerance allowed by those gages can be determined by the data given in Figure 15-4. To accomplish this it is only necessary to measure the pin or hole size of the gage, which is the virtual size in the formulas.

Most components contain several features. Therefore, a feature location gage must be constructed with a sufficient number of gaging features to gage all of the related component features in the pattern simultaneously. Although a pattern of features is shown to be acceptable according to its position tolerance requirements, the feature sizes must also be gaged or measured independently. The only exception to this requirement is that when zero position tolerances requirements are encountered, the size limit related to the critical size need not be checked since this function is performed by the feature location Gage.

MMC Holes

As seen in Figure 15-4, MMC (or clearance) holes are commonly gaged by the use of pins which use the virtual feature size for their nominal size. The pins may be either fixed to a base or inserted through the component into bushings or mounting holes in the gage. In either case, the pins must be at least as long as the holes are deep, but need not be longer. The pins should normally be the same shape as the feature; however, round pins are often used to check square holes when the angular position of the holes is not important. Paired pins may also be used to check slot locations. These techniques are illustrated in Figure 15-5.

MMC Projections and Mounting Holes

MMC projections and mounting holes are gaged in a similar manner. Projections are surrounded by their virtual size and they are, therefore, evaluated by the use of a gage plate containing holes the same shape as the component projections and sized as in Figure 15-4.

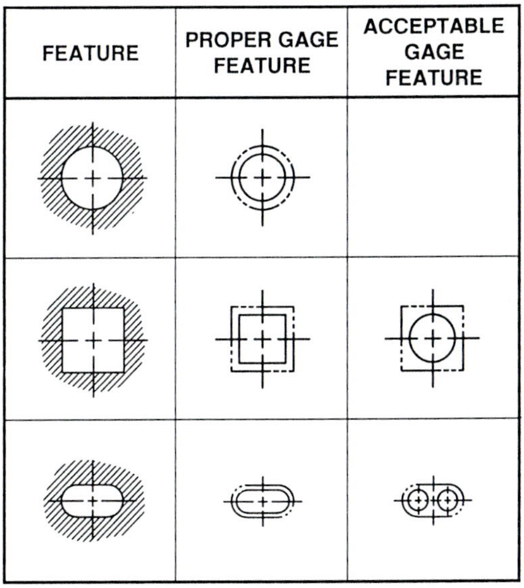

FEATURE	PROPER GAGE FEATURE	ACCEPTABLE GAGE FEATURE

Figure 15-5. Design of gage features

The *gage plate* is made equal in thickness to the height of the projection unless a projected height value is stated. When a projected height value is given, the *gage plate thickness always* assumes this value.

LMC Holes

LMC holes are similar to projections in that they are surrounded by their virtual size. This suggests that such features can be examined by use of a template similar to the gage plate described above. Although such techniques are logical, they are difficult to use and LMC holes are often checked more easily at a higher assembly level after the components they position are added. This technique was discussed in Chapter 10.

Threaded and Other Mounting Holes

The most common practice for gaging the location of threaded holes or other mounting holes is to insert pins into the holes and pass a gage plate over the pins to check the component. The gage plate is always made equal in thickness to the stated projected height value. If the pins are inserted through the plate into the part, they may be as long as, or longer than, the gage plate thickness. However, if the plate is to be passed over preassembled pins, the pins must be the same height as the gage plate thickness for the length of their gaging diameter.

If a fastener height designation is given, the pins must be equal in length to the stated fastener height, and the gage plate must be passed over the preassembled pins. If RFS holes are being gaged, the gage pins must be designed so that when they are installed in the component they centralize themselves with the axes of the holes.

For threaded holes, the thread form of the gage pin is often made to the low-dimensional limit of the internal thread form of the hole. Other prefer to make the thread form of the gage pin to the high-dimensional limit of an external thread of the same class of fit specified for the threaded hole. In any case, the portion of the pin which projects into the gage plate may be made to any convenient size so long as the proper relation between the size of the hole in the gage plate and the pin size is maintained.

Geometric Toleranced Features

Most geometric tolerances are used to determine if a surface lies within a specified tolerance zone. Generally, this is done by gaging (measuring) from a reference surface to the controlled surface. The actual gaging being done by indicators or sensors in the case of a variable gage, or by the use of feelers, flush pins, or the like in an attribute gage. Again, it is possible to design a gage that can work in both modes. Specific tolerance types are checked as follows:

Form. The part is usually placed in contact with the reference (perfect) surface and errors are measured by the deviation of the controlled surface from the reference surface. Template and feeler checks typify this type of gaging.

Attitude. In this case, an indicator or sensor tracks the reference surface, which is separated from the controlled surface. The indicator or sensor scans over the controlled surface and the range of measurements indicates the attitude variation. This type of gage is usually variable as it can be difficult to incorporate this type of check in an attribute gage.

Location. This type of gage is similar to the attitude gage except that the indicator is zeroed to a master (nominal) part. Because the variation limits are now located, an attribute gage can be constructed as an attribute gage using a maximum, or no-go, feeler and a minimum, or go, feeler to gage the separation from the controlled surface. Obviously, a variable gage can easily check both location and attitude.

Positioning the Component

The second consideration in developing a gage, after feature gaging, is the positioning of the component. If datums are specified, the part must be positioned against these surfaces prior to the gaging of the features. Hence, the same positioning devices described for fixturing may be used. If the component has piloting surfaces, the same techniques described in Figure 15-2 are employed. When composite position tolerances are encountered, the within pattern requirement is gaged without positioning the part. Then, the part is positioned to the datums given and the hole locations are again checked. This type gage will usually have removable datums and two sizes of pins for the hole location checks. Stepped pins may also be used.

Non-Rigid Parts

When non-rigid parts are encountered, they must be restrained to the specified datums prior to checks of the controlled surfaces. In some cases both free state checks and restrained checks are required. Refer to the section on non-rigid parts in Chapter 6, Datums.

Tolerances for Gage Features

The final consideration of gage design is the tolerances used to size and position the gage features. Gage features should be toleranced so that Class I errors are generated, that is, gages should err toward rejecting good parts as oppossed to accepting bad ones. Economics are less of a consideration for gage feature tolerances since inaccurate gage features cause the rejection of good parts. Thus, any savings in gage cost, resulting from more generous tolerancing, would be offset by an increase in cost due to additional rejection of good parts. Since gages are generally used only for high production volumes, small gage tolerances will usually be economical. Figure 15-6 illustrates a gage for the component described in Figure 9-17.

GAGE R & R

Gage R & R is an acronym for *gage repeatability and reproducibility.* It is the catch-all phase one encounters when one starts to examine if a gage is good enough to do its job of determining the acceptability of a part. It is the same question one encounters with any

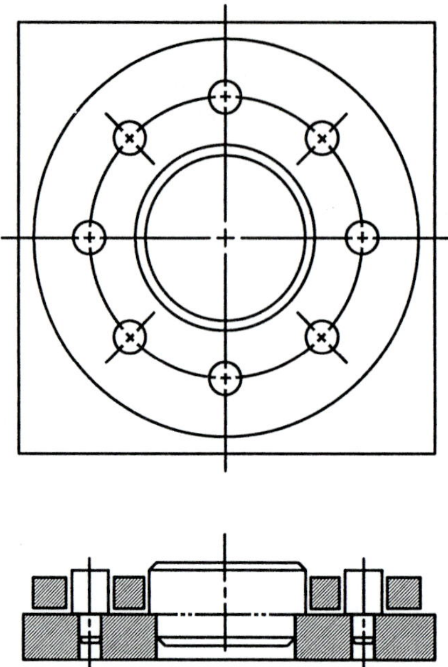

Figure 15-6. Gage for component in Fig. 9-17

measurement system . . . is the process capable of identifying the errors in the part being checked or are the system errors so large that the part errors can not be discriminated? Generally, the gage or measurement system must be able to measure less than 1/10 of the total product tolerance or process spread to be acceptable.

Variable gage or measurement system errors can be classified in several categories as shown in Figure 15-7. As in previous discussions, the errors are statistical in nature and are either related to accurate centering of the measurements or the spread of error in individual measurements.

Gage accuracy is the difference in the observed (by the gage) average measurement and the true average measurement as determined by measuring with the most accurate measuring equipment available (see Figure 15-7a)[1]. An example would be to compare the measurements of a standard *gage block* (determined by extremely accurate equipment) with its given value. Measurement equipment is certified to traceable (very accurate) standards on a periodic basis to maintain accuracy.

Gage reproducibility is the variation in average measurements made by different operators using the same gage measuring the same part (see Figure 15-7b). A classic example is the micrometer when one operator tightens it more than another.

Gage repeatability is the random variation in measurements when one operator uses the same gage to measure the same part several times (see Figure 15-7c). Other errors inducing variation in the observed average include:

[1]Figures 15-7 and 15-8, and Tables 15-1 and 15-2 are reproduced from the *Measurement Systems Analysis Reference Manual* with the permission of its publisher the AIAG (Automotive Industry Action Group).

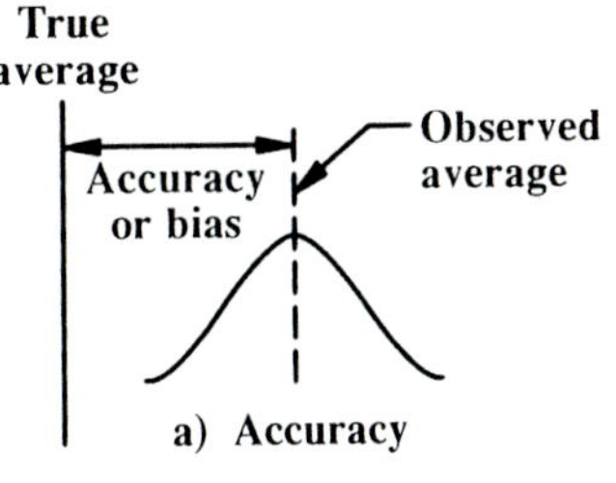

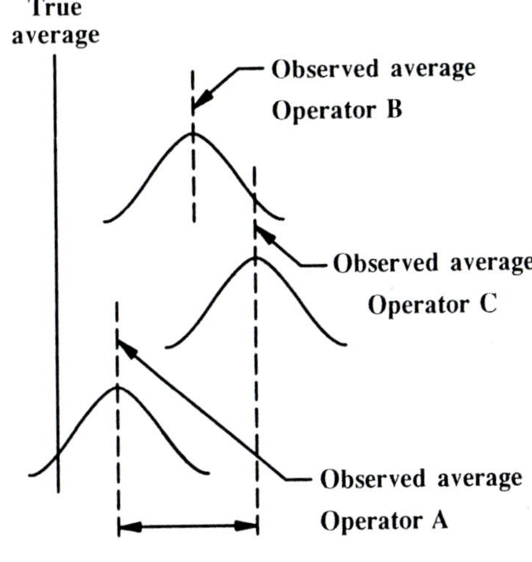

b) Reproducibility

True
Average

Observed
average

Repeatability

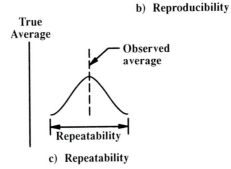

c) Repeatability

Figure 15-7. Gage Errors

Gage stability is the difference in average of two sets of measurements made at different times or temperatures (see Figure 15-8a).

Gage linearity is the difference in average of two sets of measurements made at different points in the operating range (see Figure 15-8b).

Typical Gage R & R

Tables 15-1 and 15-2 show a typical gage R & R form that was extracted from the SPC manual of a major automotive company. In conducting the study, the following steps were taken:

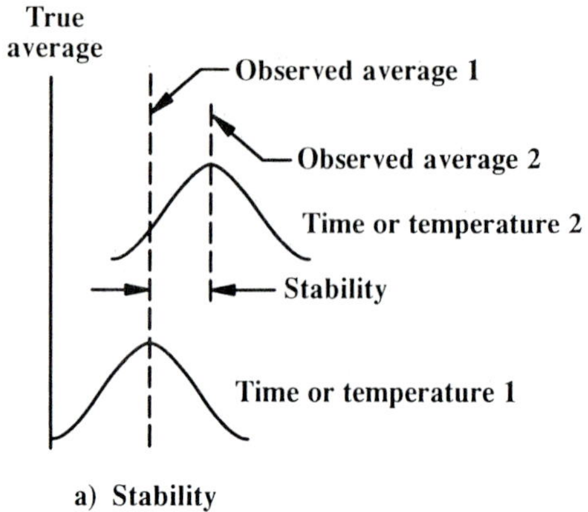

a) Stability

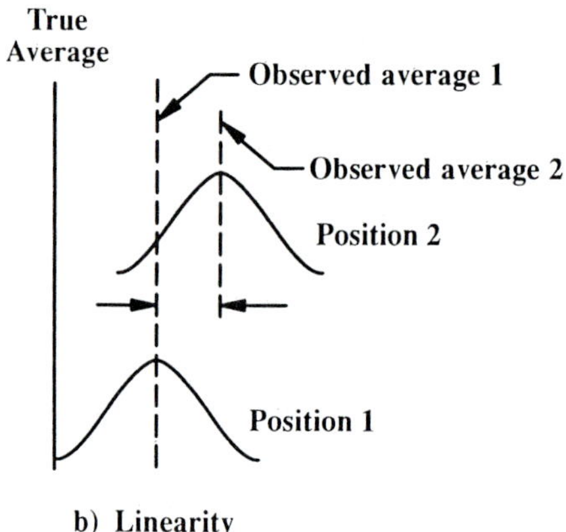

b) Linearity

Figure 15-8.

1. Number the parts 1 through 10 so that the numbers are not visible to the operators.
2. Calibrate the gage.
3. Let Operator A measure 10 parts in a random order and enter the results in Column 1.
4. Repeat steps 1 to 3 with Operators B and C.
5. Repeat the cycle, with the 10 parts measured in another random order, for the number of trials required.
6. Steps 3 through 5 may be modified for large size parts, unavailability of parts, or when operators are on different shifts.
7. Using the form shown in Tables 15-1 and 15-2, enter the observations on Table 15-1 and calculate gage R & R using the formulas shown on Table 15-2.

Results of the gage R & R study will identify if the gage is acceptable for its intended application. Whether or not a measurement system is satisfactory depends largely upon the percentage of part tolerance that is consumed by the gage system error . . . a combination of

Table 15-1.

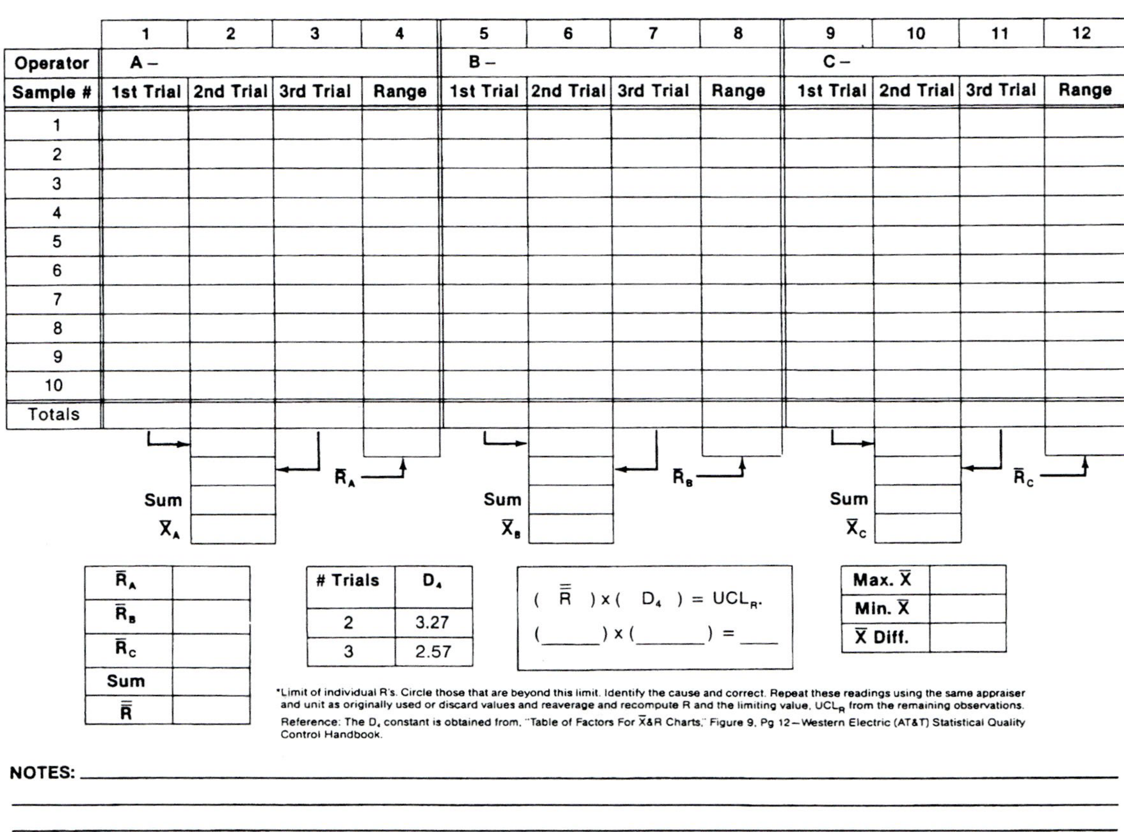

GAGE REPEATABILITY AND REPRODUCIBILITY DATA SHEET (Long Method)

	1	2	3	4	5	6	7	8	9	10	11	12
Operator	A –				B –				C –			
Sample #	1st Trial	2nd Trial	3rd Trial	Range	1st Trial	2nd Trial	3rd Trial	Range	1st Trial	2nd Trial	3rd Trial	Range
1												
2												
3												
4												
5												
6												
7												
8												
9												
10												
Totals												

	Operator A	Operator B	Operator C
Sum			
$\bar{X}$	$\bar{X}_A$	$\bar{X}_B$	$\bar{X}_C$
$\bar{R}$	$\bar{R}_A$	$\bar{R}_B$	$\bar{R}_C$

$\bar{R}_A$	
$\bar{R}_B$	
$\bar{R}_C$	
Sum	
$\bar{\bar{R}}$	

# Trials	D_4
2	3.27
3	2.57

$(\bar{\bar{R}}) \times (D_4) = UCL_R$*

(_____) x (_______) = ____

Max. $\bar{X}$	
Min. $\bar{X}$	
$\bar{X}$ Diff.	

*Limit of individual R's. Circle those that are beyond this limit. Identify the cause and correct. Repeat these readings using the same appraiser and unit as originally used or discard values and reaverage and recompute R and the limiting value, UCL_R from the remaining observations.

Reference: The D_4 constant is obtained from, "Table of Factors For $\bar{X}$&R Charts," Figure 9, Pg 12—Western Electric (AT&T) Statistical Quality Control Handbook.

NOTES: ____

Table 15-2.

Gage Repeatability and Reproducibility Report

Part No. & Name ____________ Gage Name ____________ Date ____________

Characteristic ____________ Gage No. ____________ Performed By ____________

Specification ____________ Gage Type ____________

From Data Sheet: $\bar{\bar{R}}$ = [ ] $\bar{X}_{Diff}$ = [ ]

MEASUREMENT UNIT ANALYSIS **% TOLERANCE ANALYSIS**

Repeatability—Equipment Variation (E.V.)

$E.V. = (\bar{\bar{R}}) \times (K_1)$

= ( ) X ( )

= [ ]

No. Trials (m)	2	3
K_1	4.56	3.05

% E.V. = 100 [(E.V.)/(Tolerance)]

= 100 [( )/( )]

= [ ]

Reproducibility—Appraiser Variation (A.V.)

$A.V. = \sqrt{[(\bar{X}_{diff}) \times (K_2)]^2 - [(E.V.)^2/(n \times m)]}$

$= \sqrt{[(\quad) \times (\quad)]^2 - [(\quad)^2/(\quad)]}$

= [ ]

Operators	2	3
K_2	3.65	2.70

n = number of parts
m = number of trials

% A.V. = 100 [(A.V.)/(Tolerance)]

= 100 [( )/( )]

= [ ]

Repeatability and Reproducibility (R&R)

$R\&R = \sqrt{(E.V.)^2 + (A.V.)^2}$

$= \sqrt{(\quad)^2 + (\quad)^2}$

= [ ]

$\% R\&R = 100 \left[\frac{(R\&R)}{(Tolerance)}\right]$

$= 100 \left[\frac{(\quad)}{(\quad)}\right]$

= [ ]

All calculations are based upon predicting 5.15σ (99.0% of the area under the normal distribution curve).

K_1 is $\frac{5.15}{d_2^*}$ where d_2^* is dependent on the number of trials (m) and the (number of parts (n) times the number of operators) (g) which is assumed to be greater than 15.

A.V.—if a negative value is calculated under the square root sign, the appraiser variation (A.V.) defaults to zero (0).

K_2 is also $\frac{5.15}{d_2^*}$ where d_2^* is dependent on the number of operators (m) and (g) is (1), since there is only one range calculation.

d_2^* is obtained from Table D3, "Quality Control and Industrial Statistics," A. J. Duncan.
K_1 and K_2 values were obtained from "Measurement System Analysis," General Motors GM-1729

accuracy, repeatability, reproducibility, stability, and linearity. Generally, the criteria for the acceptance of Gage repeatability and reproducibility are:

- Under 10% error—acceptable
- 10% to 30% error—may be acceptable based upon the importance of the application, gage cost and the cost of repairs, etc.
- Over 30%—generally not acceptable. Every effort to identify and correct the problem should be made.

Attribute gages can also be studied for interaction with different operators. Most quality control texts outline such methods, which usually require a large number of trials with common results between operators. Attribute gages are not usually as sensitive to operator influence as variable gages.

SUMMARY

The continuity of product function, design analysis, and drawing specification is maintained by proper design of fixtures and gages. The principal function of a fixture is to hold a part and position the tools used to form the part features. The purpose of a gage is to verify that the component being inspected meets engineering specifications. A secondary purpose is to provide data for process control. Attribute gages work by simulating the boundary conditions of the features being gaged. Variable gages use indicators or sensors to gather variable data which may be used for SPC. Gages for geometric tolerances are constructed to principles based upon form, attitude, and location control. Non-rigid parts may require restraint to perform some of the gage checks. Fixture accuracy is largely an economic issue, however, gage accuracy, particularly for variable gages, will usually be driven by R & R requirements.

Chapter 16

Inspection

As we shall use the term, *inspection* includes all the methods of product verification other than functional gaging or actual assembly, including open setup inspection using a *surface plate* and *height gage*, or inspection through the use of any of the various types of *coordinate measuring machines* that are available. Most inspection processes may be classified as variable processes in that they yield values of error rather than simply accepting or rejecting parts as do functional gages. This characteristic permits the study and further evaluation of any part that is rejected and the application of statistical methods when the sample size is sufficient.

Inspection techniques are valuable when verification is required for purchased components, experimental parts, small lots of production parts, or in other circumstances where the construction of gages would not be economical. There is a widely held misconception that the use of GD&T necessitates verification by the use of gages. In reality, only parts with sufficiently high-volume or high-inspection costs justify the cost of special gages.

This section outlines several methods which can be used to inspect parts when gages are not available or when it is not economical to construct or purchase them. Techniques and shortcuts which can be introduced into an inspection are discussed.

BASIC TECHNIQUES—POSITION TOLERANCE

There are three key steps to any inspection of feature locations:

1. Determine the datum or starting point for measurement.
2. Measure the feature sizes and locations and record the measurements.
3. Determine the allowable position tolerance for each feature, when necessary.

Selection of a Datum

Determination of the datum system allows a part to be positioned for measurement. If GD&T has been used, all the necessary datums will be specified. If GD&T has not been used, the datum selection will require careful consideration.

Current standards recognize two types of position tolerance relationships to datums:

1. *Pattern location tolerances* located directly from datums; or
2. *Within pattern tolerances* which are freed from all except the primary datum. This is found with composite position tolerance specifications.

The first case is easily handled whether it is a stand alone requirement or the pattern location portion of a composite position tolerance specification as measurements from the datums can be compared directly to the specifications. The within pattern requirements of a composite position tolerance specification require special consideration as the location errors must be freed from the measurement datums to complete the evaluation.

Measurement of Feature Locations

The next step is to measure the feature sizes and locations and record the measurement data. The type of position tolerance specified influences the manner in which the feature locations are measured. For clearance holes, the tolerance applies throughout the depth of the hole, and measurements should be made over as great a length as possible. In the case of mounting holes, the measurements are made about the surface and should extend over the length of the specified projected height. Such measurements are made by using pins placed into the holes so that they project above the surface of the part. In the case of threaded holes, thread gages can be inserted into the holes and measurements taken from the shanks of the gages. Projections should be checked over their full height.

It is not easy to obtain measurements over the full length of a deep clearance hole, nor is it always easy to fit pins to mounting holes so that they are sufficiently stable to permit accurate measurements to be taken. Further, when the holes are accessible and measurements are obtained at extreme positions, it is difficult to reconcile differing values when determining the acceptability of a part. Consequently, inspections usually necessitate certain reasonable approximations. For example, although a hole tapers, it may be assumed to possess its minimum size over its full length. Hole locations may be checked only at the surface of the part, or, when checked over some length, an extreme value may be used as a single location coordinate.

Since most measurement systems are based upon coordinate systems, the deviations from true position must be converted to diametral measure for round holes. These conversions are made using the techniques shown in Chapter 11.

Determination of Allowable Tolerance

Using the measured feature sizes and the position tolerance specification, the allowable position tolerance for each feature can be determined through the use of the methods suggested in Chapter 9 when required.

OPEN SETUP INSPECTION

The *open setup* type of inspection is usually made on a surface plate using a right angle plate and an indicating height gage. Measurements are made to the surface of a hole, or over the diameter of a fitted pin or projection. To obtain center distance dimensions, the radius of the pin must be added to, or subtracted from, the measurements to the hole or pin surfaces. After all the feature locations are measured in the vertical direction, the right angle plate is placed on its side and the horizontal locations of the features are measured in the vertical direction. This procedure is depicted in Figure 16-1. Turning the plate from the horizontal to the vertical, and the necessary calculations, contribute to the potential for error.

Inspection Examples

Example 16-1 illustrates a position tolerance specification and the results of an open setup inspection. Notice that an inspection form is used and that the datums, as well as the hole sizes and locations, are recorded on the form. Such a practice is highly recommended because it eliminates the necessity of measuring the part a second time if it should be rejected, or other questions arise. It should also be noted that some of the calculations can be performed by the computer of many coordinate measuring machines (CMM). The calculations are carried out completely here for instructional purposes and inspection procedures can be abbreviated where such capability exists.

In Example 16-1, rectangular coordinates are converted to diametral measure using a hand-held calculator and the formula of Chapter 11. The diametral error values are compared to the position tolerance and found to be acceptable, hence additional tolerance values are not calculated. Example 16-2 details the inspection of a similar part with 4 projections. Two of the 4 locations require calculation of the additional tolerance to accept the part. Example 16-3 illustrates the inspection of a part containing square holes. Again the techniques used are the same as in the first two examples, the point illustrated being that the tolerances must be examined in two directions.

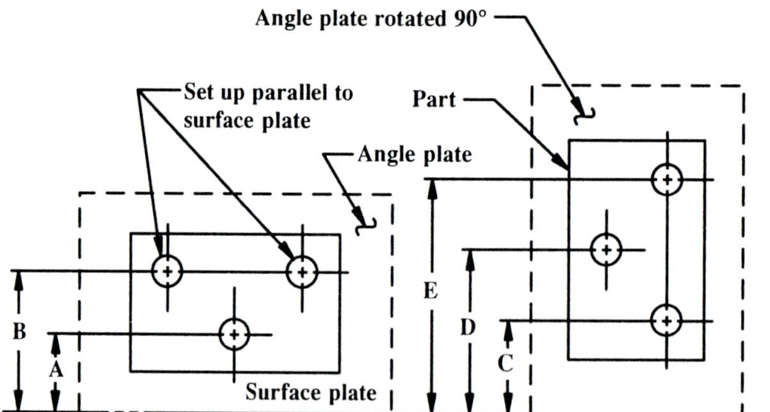

Figure 16-1. Typical inspection procedure. In this procedure dimensions A and B are measured, the angle plate is turned 90 degrees, and dimensions C, D, and E are then measured

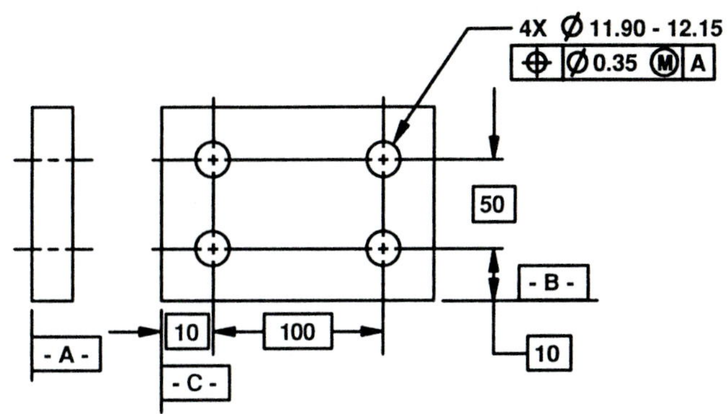

Variation Management Consultants
Northville, Michigan
Inspection Report

Part No.:

Date: 3/13/92 **Insp.** J.V.L.

Sketch:

+1 2+
+3 4+

∧ = Datum

Hole	x	y
1	10.16	60.04
2	109.94	59.88
3	10.10	9.96
4	110.13	10.06

Notes:

Hole Sizes
1 – 11.92
2 – 11.85
3 – 11.94
4 – 12.05

All O.K.

Calculation:

Hole 1 x = 0.16 y = 0.04

$\phi_1 = 0.330 <$ ⌖ Ø 0.35

Hole 2 x = 0.06 y = 0.12

$\phi_2 = 0.268 <$ ⌖ Ø 0.35

Hole 3 x = 0.10 y = 0.04

$\phi_3 = 0.215 <$ ⌖ Ø 0.35

Hole 4 x = 0.13 y = 0.06

$\phi_4 = 0.286 <$ ⌖ Ø 0.35

Part accepted

Example 16-1.

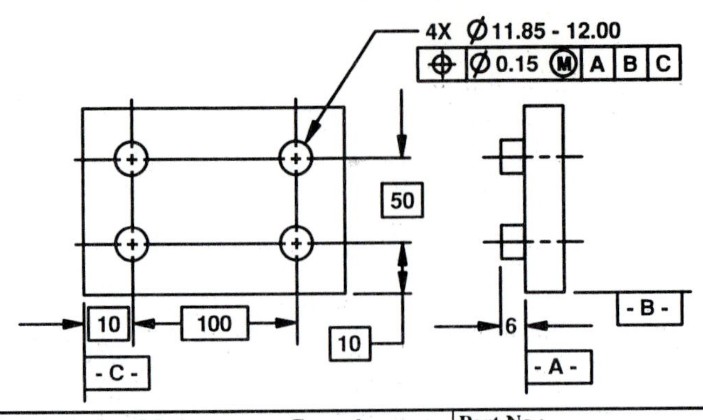

Variation Management Consultants
Northville, Michigan
Inspection Report

Part No.:

Date: 3/13/92 | Insp. J.V.L.

Sketch:

$+^1$ $^2+$

$+_3$ $_4+$

Λ = Datum

Proj.	x	y
1	10.02	60.10
2	109.90	59.92
3	10.05	9.95
4	110.05	10.03

Notes:

Projection Sizes
1 – 11.92
2 – 11.87
3 – 11.92
4 – 11.90

All within size limits

Calculation:

Proj. 1 $x = 0.02 \quad y = 0.10$

$\varnothing_1 = 0.204 >$ ⌖ $\varnothing\ 0.15 < T_A$ below

$T_{ADD} = 12.00 - 11.92 = 0.08$

$T_A = 0.15 + 0.08 = 0.23$

Proj. 2 $x = 0.10 \quad y = 0.08$

$\varnothing_2 = 0.256 >$ ⌖ $\varnothing\ 0.15 < T_A$ below

$T_{ADD} = 12.00 - 11.87 = 0.13$

$T_A = 0.15 + 0.13 = 0.28$

Proj. 3 $x = 0.05 \quad y = 0.05$

$\varnothing_3 = 0.142 <$ ⌖ $\varnothing\ 0.15$

Proj. 4 $x = 0.05 \quad y = 0.03$

$\varnothing_4 = 0.117 <$ ⌖ $\varnothing\ 0.15$

Part accepted

Example 16-2.

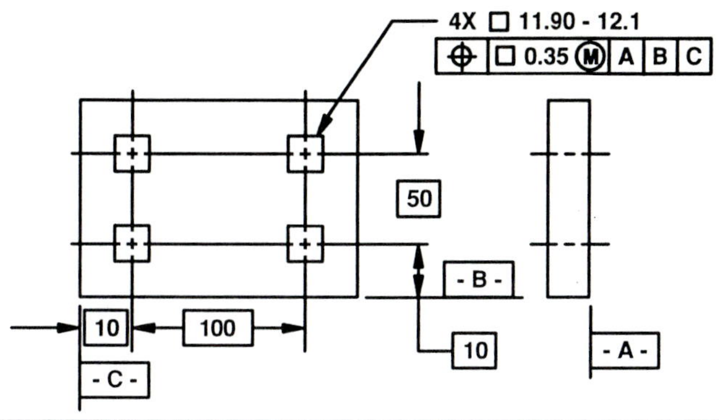

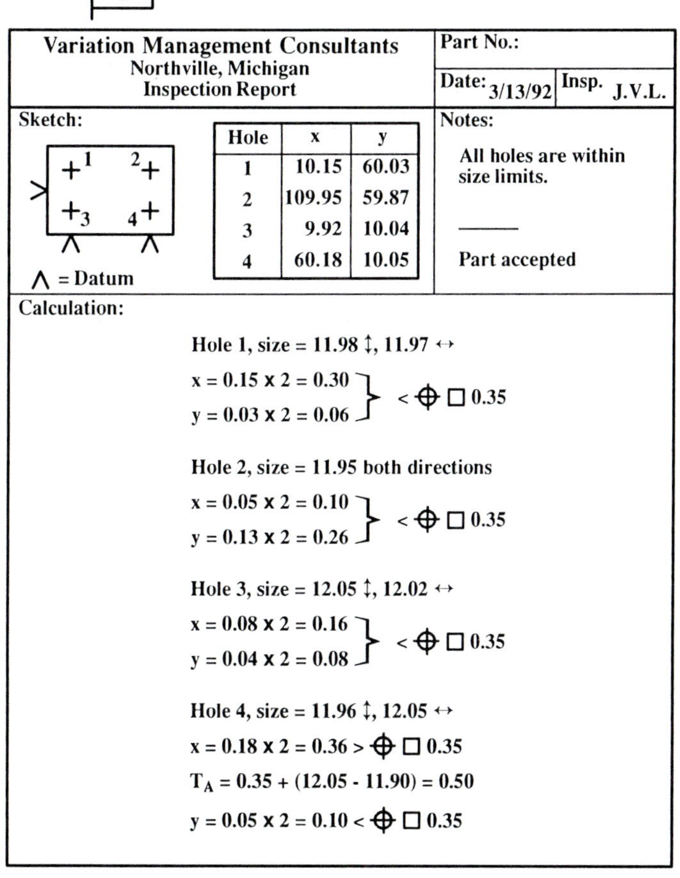

Variation Management Consultants
Northville, Michigan
Inspection Report

Part No.:

Date: 3/13/92 | Insp. J.V.L.

Sketch:

Hole	x	y
1	10.15	60.03
2	109.95	59.87
3	9.92	10.04
4	60.18	10.05

Notes:

All holes are within size limits.

———

Part accepted

Calculation:

Hole 1, size = 11.98 ↕, 11.97 ↔

x = 0.15 x 2 = 0.30
y = 0.03 x 2 = 0.06 } < ⌖ □ 0.35

Hole 2, size = 11.95 both directions

x = 0.05 x 2 = 0.10
y = 0.13 x 2 = 0.26 } < ⌖ □ 0.35

Hole 3, size = 12.05 ↕, 12.02 ↔

x = 0.08 x 2 = 0.16
y = 0.04 x 2 = 0.08 } < ⌖ □ 0.35

Hole 4, size = 11.96 ↕, 12.05 ↔

x = 0.18 x 2 = 0.36 > ⌖ □ 0.35

T_A = 0.35 + (12.05 - 11.90) = 0.50

y = 0.05 x 2 = 0.10 < ⌖ □ 0.35

Example 16-3.

PATTERN LOCATION TOLERANCES

When the placement of a hole pattern is defined by composite position tolerances (as discussed in Chapter 6), inspection will require other techniques. The pattern location tolerances are easily evaluated (see the figure in Example 16-4). However, it is easily seen that the individual location errors all exceed the within pattern tolerance. You must then free the location error pattern from the datums to see if the within pattern requirements can be met. Basically, this requires iteration as shown in Example 16-5, the use of CMM computer-based iteration, or the technique shown in Chapter 17. Example 16-5 shows that the part is acceptable by a trial and error method.

SUMMARY OF INSPECTION EVALUATION PROCEDURE

Based upon the foregoing discussion and examples, the inspection procedure for position tolerances can be summarized as follows:

1. Measure all feature sizes. If feature sizes are out of tolerance, the part can be rejected at this point.
2. Establish the datums and position the part for measurement.
3. Measure all feature locations, establish coordinate deviations, and convert to diametral measure, if necessary.
4. Compare feature location deviations to the specified position tolerance. If acceptable, the inspection is complete.
5. If the part is not acceptable at step 4, calculate the additional position tolerance resulting from favorable feature size. Correct the specified position tolerance to determine an allowable position tolerance.
6. Compare feature location deviations to the allowable position tolerance. If acceptable, the inspection is complete.
7. If the part fails to meet within pattern location requirements based upon measurements from datums, iterative techniques can be used to complete the evaluation.

BASIC TECHNIQUES—GEOMETRIC TOLERANCES

As almost all *geometric tolerances* are surface requirements, inspection should nominally be a surface check. Since most inspection techniques, including coordinate measuring machines (CMM) yield point measurements, inspection of geometrically toleranced surfaces is basically a compromise. Unfortunately, such systems suffer from a class II error (i.e., they err toward accepting bad parts since some potentially bad points may not be measured). As the point data cannot fully qualify a surface, the quality and quantity of the points selected directly affect the quality of the compromised observation. The author anticipates that rapid scanning mechanisms will ultimately be interfaced with CMMs to more fully qualify the total area of the measured surface to yield a higher quality measurement.

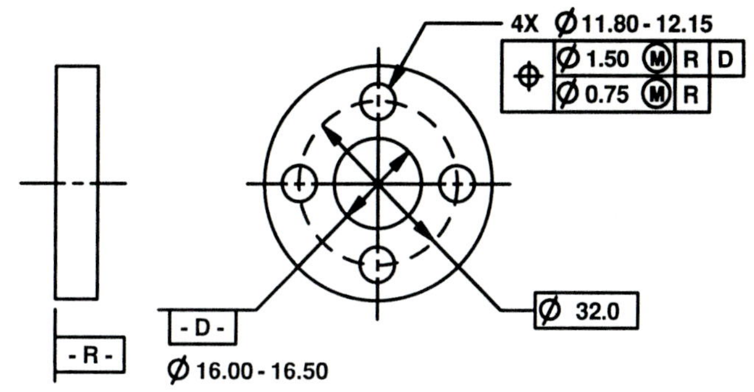

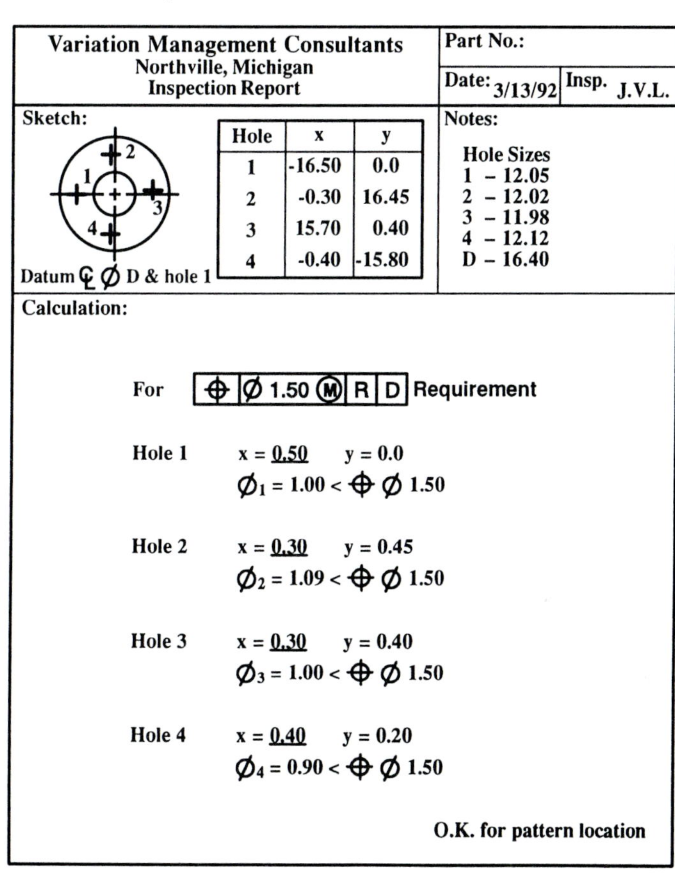

Variation Management Consultants
Northville, Michigan
Inspection Report

Part No.:

Date: 3/13/92 Insp. J.V.L.

Sketch:

Datum ℄ ⌀ D & hole 1

Hole	x	y
1	-16.50	0.0
2	-0.30	16.45
3	15.70	0.40
4	-0.40	-15.80

Notes:

Hole Sizes
1 – 12.05
2 – 12.02
3 – 11.98
4 – 12.12
D – 16.40

Calculation:

For ⌖ ⌀ 1.50 Ⓜ R D Requirement

Hole 1 x = 0.50 y = 0.0
$\phi_1 = 1.00 <$ ⌖ ⌀ 1.50

Hole 2 x = 0.30 y = 0.45
$\phi_2 = 1.09 <$ ⌖ ⌀ 1.50

Hole 3 x = 0.30 y = 0.40
$\phi_3 = 1.00 <$ ⌖ ⌀ 1.50

Hole 4 x = 0.40 y = 0.20
$\phi_4 = 0.90 <$ ⌖ ⌀ 1.50

O.K. for pattern location

Example 16-4.

Example 16-5

The location errors of Example 16-4 are typically high and to the left.

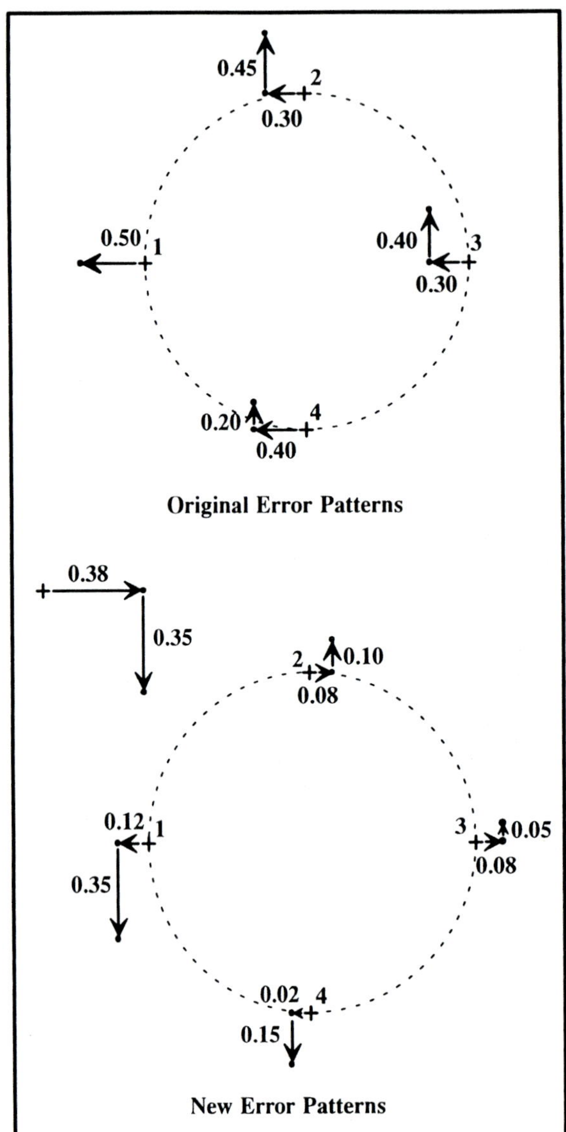

The average "up" error is 0.35 and the average "left" error is 0.38. Therefore as a first iteration, let's move the pattern to the right and down by these average values, resulting in new error values of:

Hole	*X*	*Y*	∅
1	.12	.35	.74
2	.08	.10	.26
3	.08	.05	.19
4	.02	.15	.31

We were fortunate; our first iteration shows that the within pattern requirements are met. Had this iteration proved unacceptable, two options remained:

1. Determine if additional tolerance from favorably sized features solved the problem; or
2. Try another iteration.

Location Tolerances

Inspection techniques work best in checking locations by comparing measured locations from datums to specified nominal values. The extensive use of CMMs is built upon a foundation of measuring location tolerances. *Location tolerance* inspection can be summarized as:

1. Measure specified points.
2. Calculate deviation perpendicular to nominal (most CMMs perform this operation).
3. Compare to one-half of the tolerance value (a location profile of a surface tolerance is usually centered on nominal).

Attitude Tolerance

Attitude tolerance is also easy to evaluate using point measurements; again being affected by the quality and quantity of points measured. Attitude tolerance inspection can be summarized as:

1. Measure specified points (data from a location measurement may be used, if available).
2. Sum the maximum + deviation and the maximum – deviation to determine the attitude error band.
3. Compare to specified (total) tolerance.

Example 16-6 shows how the measured deviations at 5 points are analyzed for the door edge specification of Figure 7-21.

Form Tolerance

Form tolerance is very difficult to evaluate by inspection measurements. This is because all the data acquired relates to datums, while form tolerances have no datums. Hence, it is necessary to calculate or simulate a best fit mean surface for the data and subsequently

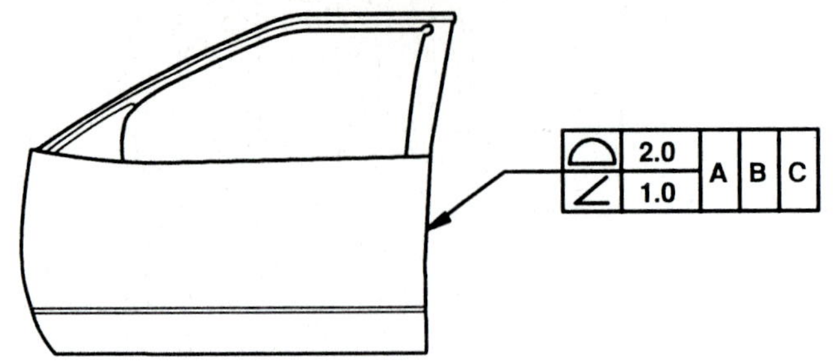

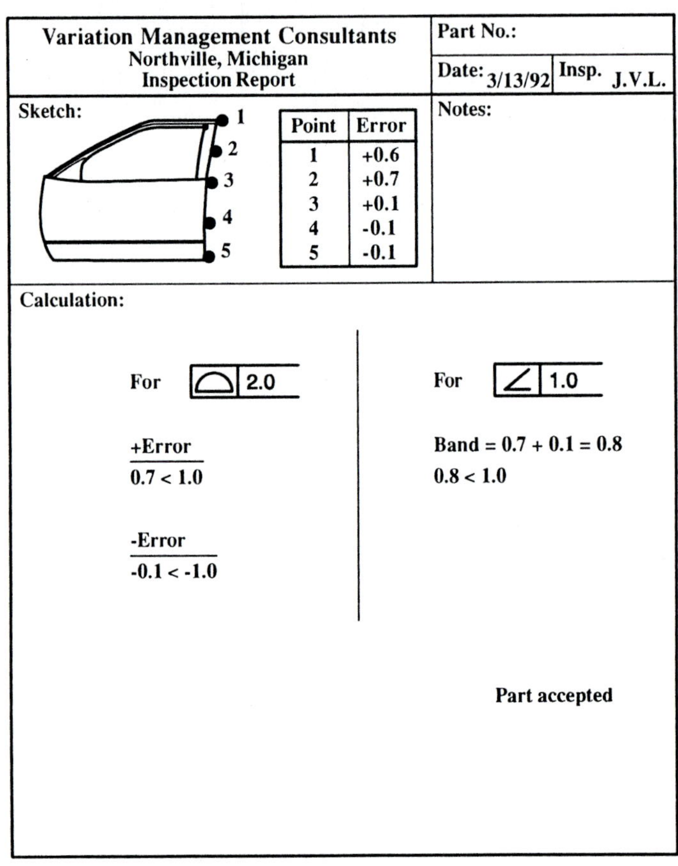

Variation Management Consultants
Northville, Michigan
Inspection Report

Part No.:

Date: 3/13/92 Insp. J.V.L.

Sketch:

Point	Error
1	+0.6
2	+0.7
3	+0.1
4	-0.1
5	-0.1

Notes:

Calculation:

For [profile of a surface] 2.0

+Error
0.7 < 1.0

-Error
-0.1 < -1.0

For [angularity] 1.0

Band = 0.7 + 0.1 = 0.8
0.8 < 1.0

Part accepted

Example 16-6.

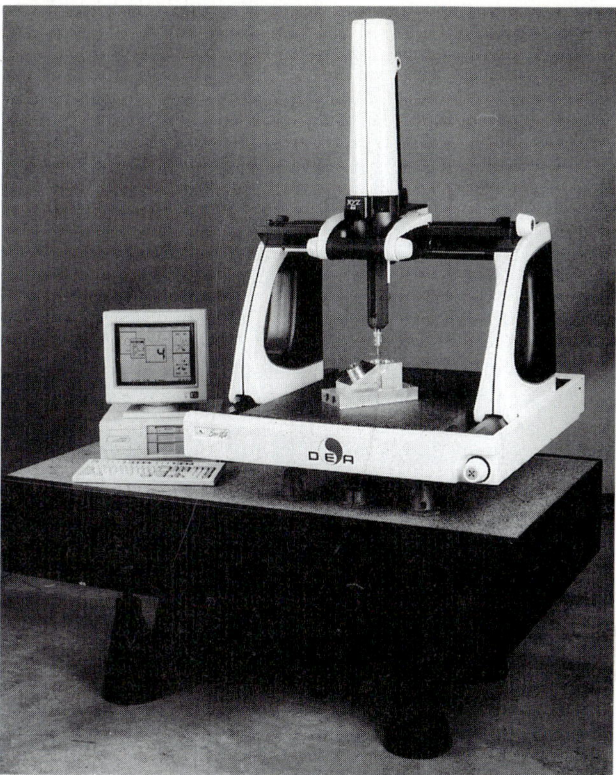

Figure 16-2. "Desk Top" CMM. (Reprinted with permission from Digital Electronic Automation, Inc., Livonia, MI.)

determine the spread of data from the best fit surface. The spread is then compared to the specified tolerance. As an example, *flatness* is determined as follows:

1. Measure several points on the surface.
2. Find a mean plane that fits the data so that errors from the plane are at a minimum. This is done by a linear regression calculation on a CMM, or remote, computer.
3. Determine the error spread from the best fit plane and compare it to the flatness tolerance.

While linear regression based upon a flat surface is fairly straightforward, similar applications to complex curved surfaces are not. Here, the nominal surface is no longer described by a formula but frequently by math data residing in a CAD computer system. While the concept described above is still applicable, the implementation is much more difficult. Computer-based simulation appears to be the only method of evaluating such requirements. Some of the better CMMs are capable of performing the necessary computer iterations. However, you must also consider if such measurement methods are appropriate,

Figure 16-3. Large CMM measuring engine block. Photo courtesy of IMT Division, Carl Zeiss, Inc.

or if gages or templates might be more appropriately used. When the type of part permits, form tolerance can be viewed with an optical comparator.

Unit basis form tolerances are even more difficult to evaluate by discrete measurements. As previously discussed, they are form tolerances applied over small lengths or areas. As such the inspection process is identical to other form tolerances except that a series of overlapping evaluations must be made for the length or area specified. Hence, unit basis form tolerances on complex curved surfaces have the same complexity as other form tolerance computations plus the need to perform the check on a limited length or area which is then moved in an iterative fashion over the controlled surface. Such tolerances and related inspection measurement techniques should be avoided, when possible.

Number of Measured Points

The *number of points measured* has a significant effect on the observed variation due to the ability (or inability) of the measurements to describe the controlled surface. Table 16-1 illustrates how a minimum number was established for certain automotive sheet metal components. Note that these minimums are a long way from an ideal description of a surface,

Table 16-1. Number of points to be measured.

GEOMETRY	TOLERANCE	MINIMUM # POINTS	LOGIC
HEM	⌓ LOCATION	2	- ASSURES PLAN VIEW CONTOUR IS CORRECT - ONE POINT WON'T DETECT ROTATION
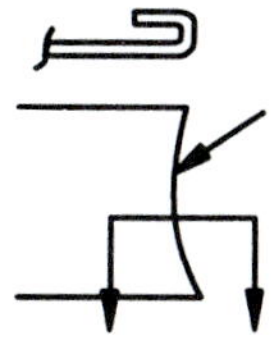	// ⊥ ∠ ATTITUDE	2	- SAME AS ABOVE
	▱ ⌓ FORM	3	- NEEDS THIRD POINT TO DEFINE CURVE
FLANGE	⌓ LOCATION	2	- ASSUMES PLAN VIEW CONTOUR IS CORRECT - IF SPRING BACK IS NOT AN ISSUE-- CHECK AT A - ONE POINT WON'T DETECT ROTATION
		4	- IF SPRING BACK IS AN ISSUE -- CHECK AT A-B
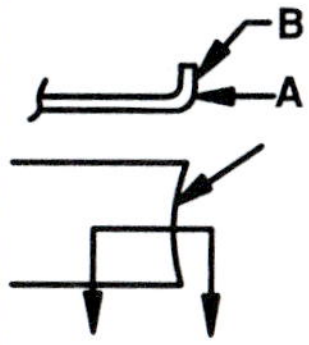	// ⊥ ∠ ATTITUDE	2	- SAME AS ABOVE
		4	
	▱ ⌓ FORM	3	- NEEDS THIRD POINT TO DEFINE CURVE
		6	- IF SPRINGBACK AFFECTS FORM
LARGE PANEL SURFACE 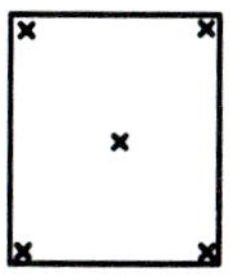	▱ ⌓ FORM	5	- FOUR CORNERS AND CENTER DETECTS WARP, TWIST OR CONCAVE/ CONVEX ERRORS

Figure 16-4. Measurement robot integrated into manufacturing. (Reprinted with permission from Digital Electronic Automation, Inc., Livonia, MI.)

but rather indicate how points and their locations may be selected and possibly standardized for specific types of parts.

COORDINATE MEASURING MACHINES

Coordinate measuring machines (CMM) now dominate the measurement laboratories of major corporations. Because of the speed with which they can take measurements and the ease with which data can be processed and displayed, or analyzed by companion computers, older setup methods often are no longer practical. Some units are programmable so that the measuring process is completed automatically once the part is positioned in the machine.

CMMs are available in a wide range of sizes from the desk top unit of Figure 16-2 to large units as in Figure 16-3. Many special applications are also possible using a basic machine or its core components. Figure 16-4 shows a robot measurement unit used on a manufacturing line for a precision component.

When CMMs are used for routine inspection of large components, especially those that lack complete rigidity, a checking (or holding) fixture is often built to support the part being inspected. Such fixtures are subject to R & R requirements as discussed in Chapter 15.

SUMMARY

Inspection, as we have used the term, includes all the methods of product verification other than functional gaging, or actual assembly. The basis inspection technique is the open set up procedure which has two steps:

1. Measurement of appropriate surfaces; and
2. Analysis of the data.

Position tolerances are fairly straight forward in their inspection. Steps include:

1. Identification of the datum;
2. Measurement of feature size and location; and
3. Calculation of allowable tolerance, when needed.

Position tolerance specifications are relatively easy to analyze if they are tied directly to datums, however, within pattern requirements associated with composite position tolerance requirements may require iterative techniques.

Geometric tolerances are more difficult to measure and analyze due to the inherent compromise resulting from measuring points to qualify surfaces. Ultimately, scanning devices may replace the discrete measurement devices. Within this category, location tolerances are the easiest to measure and analyze. Attitude tolerances can be analyzed using location tolerance data. Form tolerances are generally more difficult and inspection of these tolerances will generally be limited by the capability of the CMM support computer to analyze the acquired discrete data, hence other methods may prove to be more appropriate.

The many sizes and types of CMMs available make measurement and data analysis much faster and more accurate than the conceptually similar setup methods.

Chapter 17

Paper Gaging

Paper gaging is an inspection technique which allows inspection data to be evaluated in the same way that a composite gage verifies a part. It is one of the techniques that can be used when within pattern requirements are encountered. This technique can be used as the primary method of evaluating data from an inspection setup, or it can be used to reevaluate parts rejected by a evaluation in which within pattern requirements were tied to datums. And finally, this technique can be the basis of a computer based simulation of a composite gage.

PAPER GAGING

The paper gaging technique is illustrated by discussion of the example part shown in Figure 17-1.

Based upon the actual hole sizes, the specified position tolerance can be corrected to determine an allowable position tolerance for each hole. Then tolerance zones representing these values can be drawn to some enlarged scale, as indicated in Figure 17-2.

Next, using data from the inspection process, the actual hole locations are laid out on a sheet of transparent paper as shown in Figure 17-3. This layout is made to the same enlarged scale as the tolerance zone layout.

The final step is to superimpose the hole location layout upon the tolerance zone layout as in Figure 17-4. If the layouts can be positioned so that all of the hole centers fall within their respective tolerance zones, the conditions of the position tolerance specification are met and the part is acceptable.

The principal disadvantages of this method are the time required to make the neces-

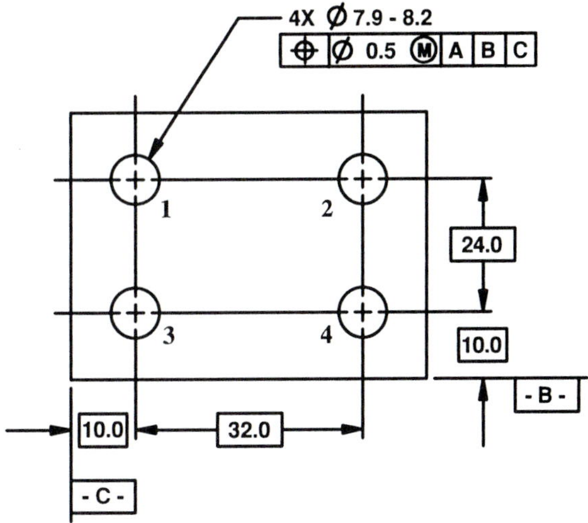

Figure 17-1. Component used to describe paper gage methods

sary layouts and the large size of the layout which results from even moderately sized hole patterns. It remains, however, an important technique an that it can be applied to any hole or feature pattern, and is completely free of limitations other than the accuracy of the layouts.

MODIFIED PAPER GAGE

The *modified paper gage* had been developed to eliminate the undesirable features of the conventional paper gage. The basis of this modification is the superposition of the several feature positions unto a common position. Consider, for example, the tolerance zone layout

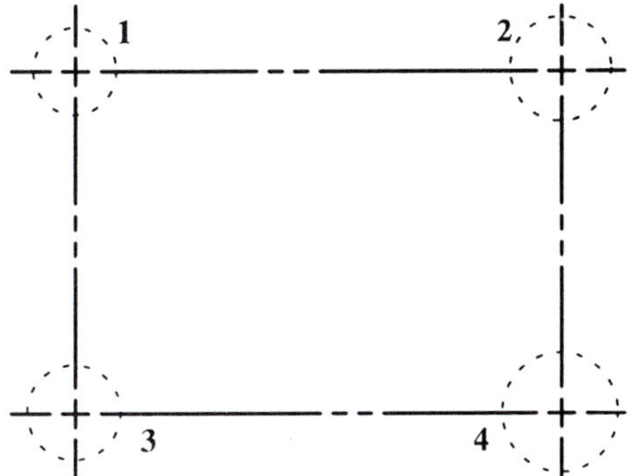

Figure 17-2. Enlarged layout of tolerance zones for example part (drawn out-of-scale for clarity)

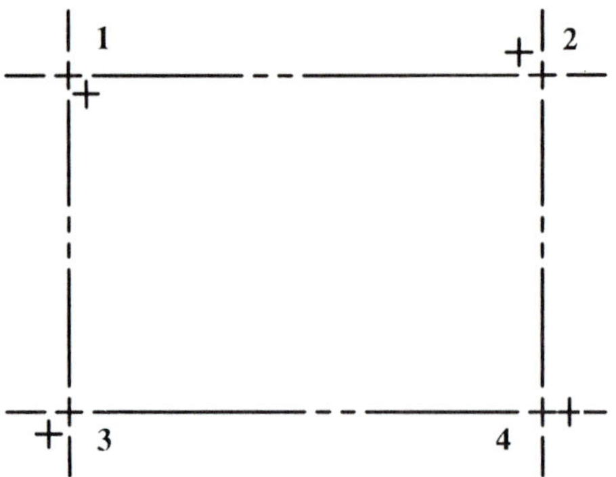

Figure 17-3. Enlarged layout of actual hole locations of example part (drawn out-of-scale for clarity)

shown in Figure 17-2. If the center of tolerance zones 1, 2, and 4 are moved horizontally and vertically by an amount equal to their nominal (or true position) location dimension from the center of zone 3, all of the zones will form concentric limits about a common reference point, Point 3, as shown in Figure 17-5.

Then the actual locations of holes 1, 2, and 4, determined by the inspection process and shown in Figure 17-3, are moved in a similar fashion to form a cluster of points about a common reference point, the nominal location of Point 3 in this case. This is accomplished in the following manner. First, consider one of the actual locations, such as Location 2 of Figure 17-3, and it true position center. Visualize the two points as marked upon a piece of transparent paper that can be easily moved. Now, move the true position location of Point 2 to the left and down by the amount of its nominal (true position) location dimension. The true position center will now lie on the common reference point (nominal Point

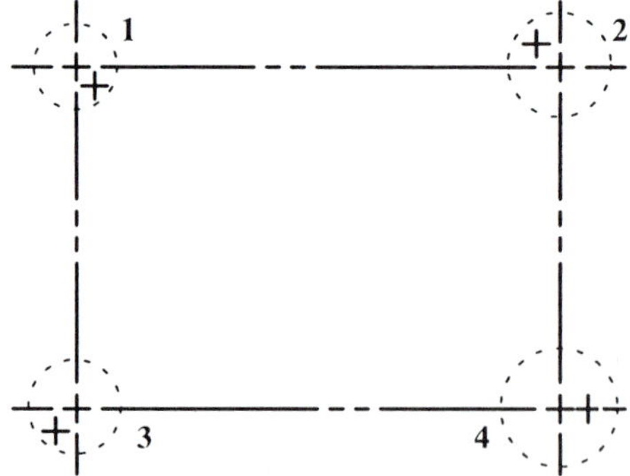

Figure 17-4. Superimposed Figs. 17-2 and 17-3 form paper gage showing acceptable part (drawn out-of-scale for clarity)

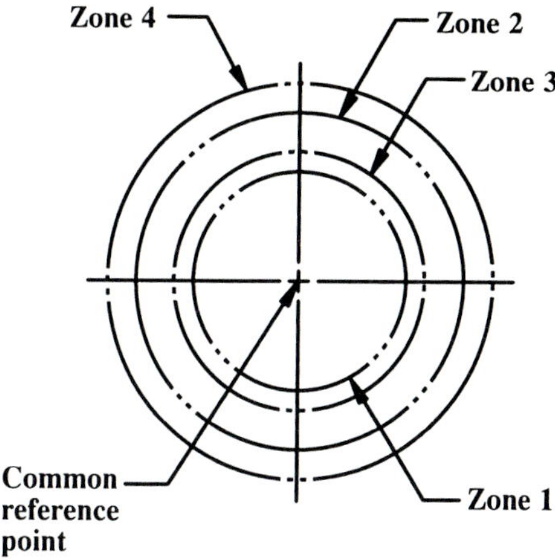

Figure 17-5. Tolerance zones of Fig. 17-2 superimposed upon a common position

3) and the actual hole location will maintain its location in the same quadrant and at the same distance from the true position because the actual location point and the true position point were moved as a pair. When the process is repeated for Points 1 and 4, the results shown in Figure 17-6 are obtained.

In actual practice the process need not be so involved. The tolerance zone sizes are simply determined and drawn about a common center on one layout, such as Figure 17-5. Since the hole location deviations from the reference point are equal to their deviation from true position, these true position deviations are determined from the inspection data and laid out from a common point on a second layout.

Since we now have two layouts in which both tolerance zones and location have been moved by the same amount, we may compare them to determine if the tolerance specifica-

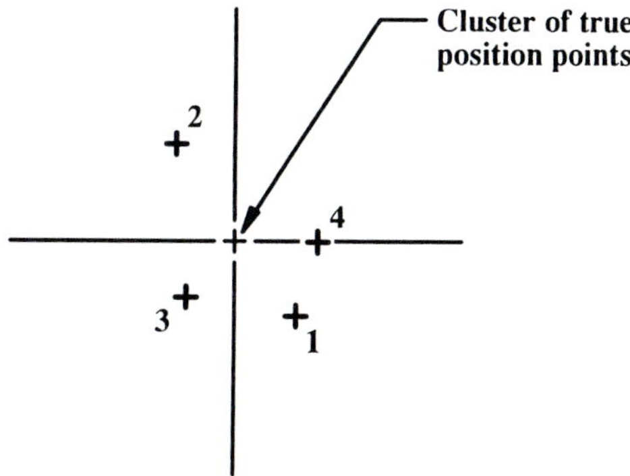

Figure 17-6. Hole locations of Fig. 17-3 superimposed upon a common position

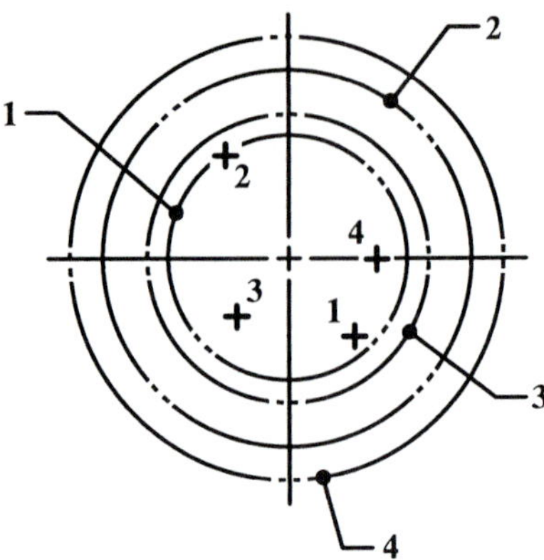

Figure 17-7. Superimposed Figs. 17-5 and 17-6 form a modified paper gage and indicate an acceptable part

tions are met. The two layouts are again superimposed, and if they can be adjusted so that each point representing a hole location is contained within its respective tolerance zone, the part is acceptable. The result of superimposing Figures 17-5 and 17-6 is shown in Figure 17-7. Note that as a consequence of eliminating the nominal dimensions, the layouts can now be made to an extremely large scale, considerably decreasing any error causes by inaccuracy of the layout work, while remaining reasonable in size. Further, any point may be selected as the reference point, or measurement datum, and the resulting layouts will be the same.

Where a datum is given in the tolerance specification, however, it should be used as the reference point for all measurements, and in such cases, the reference points on the two layouts must be aligned or contained in an appropriate tolerance zone if the datum is a feature of size.

Source of Error

One *source of error* can be introduced by the modified paper gage system. This error occurs when the part is positioned during inspection so that the location measurements are not made parallel and perpendicular to the lines of true position. When this happens, the measured deviations are no longer equal to the deviations from true position. Figure 17-8a shows a pattern of holes and two possible reference orientations during inspection. Tolerance zones are also shown to illustrate that the hole pattern is functional. In Figure 17-8b, a modified paper gage is shown for the inspection where the measurements were made parallel and perpendicular to the lines of true position. In this case it can be seen that the hole locations maintain the same position within the tolerance zones in both illustrations. Figure 17-8c represents the modified paper gage resulting from the second orientation shown in Figure 17-8a. Note that the rotation of the reference line from the line of true position has

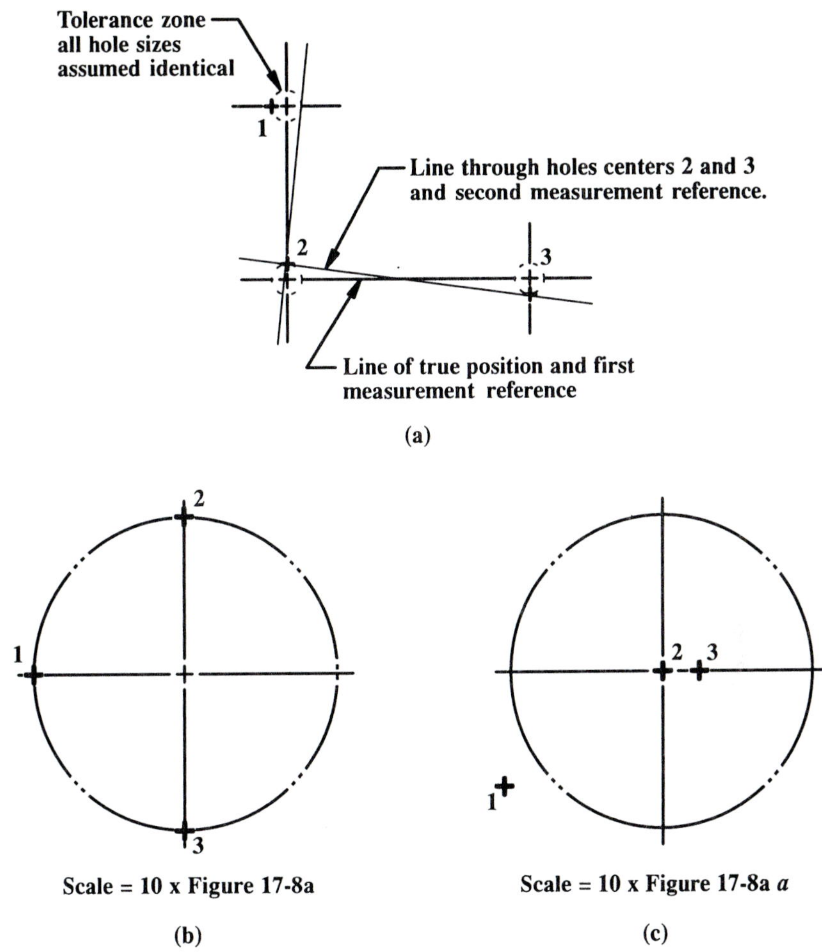

Figure 17-8. (a) Standard paper gage of functional part and alternate measurement references. Very large tolerance values are used to emphasize effect of Figs. 17-8b and 17-8c. (b) Modified paper gage from the first measurement reference shows part to be acceptable. (c) Modified paper gage from the second measurement reference shows part to be unacceptable

caused a change in the magnitude of the plotted deviations causing the modified paper gage to show the functional part to be unacceptable.

Part Orientation

It is obvious that *part orientation* becomes an important step in the inspection process when the modified paper gage system is to be used. Strict adherence to specified datums in both manufacturing and inspections operations will normally preclude such errors. However, this can be difficult to accomplish, especially with round or other symmetrical parts where datum identification can be a problem.

In addition, cases will arise in which it is desirable to use the modified paper gage system when the tooling procedure is not known, as in the case of purchased components. Certain observations permit such situations to be evaluated in an intelligent manner. If Figure 17-8 is reexamined, it can be seen that the rotation of the line of orientation created a class I error, causing a functional part to appear as unacceptable due to the enlargement of the measured deviations. Further consideration of this fact will indicate that an undesirable orientation cannot make a nonfunctional part appear acceptable. Because of this important conclusion, it is always permissible to use the modified paper gage system, whether or not proper orientation can be assured. When a part appears unacceptable, and the inspection orientation is questionable, conventional paper gage layouts may be made to obtain an exact evaluation of the part, or the conventional paper gage may simply be sketched, a new set of reference lines assumed, and new tolerance deviations estimated. These estimated deviations are then analyzed on a modified paper gage to determine whether the component is acceptable. With practice, this procedure can be performed quickly and accurately.

STANDARDIZED PAPER GAGE FORMS

A real advantage of the paper gage system is that a *standardized position grid* and *tolerance zone overlay* can be used, significantly reducing the time required to evaluate inspection data. The tolerance zones are represented by concentric circles drawn on a transparent overlay with the tolerance zone sizes labeled on the circle. The position of the holes are plotted

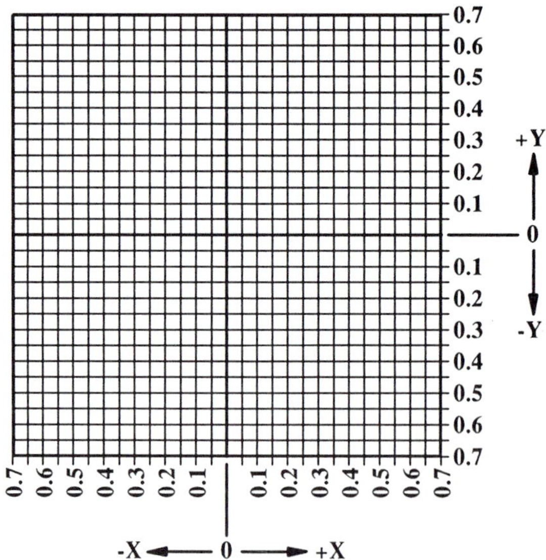

Figure 17-9. Standardized position grid for modified paper gage analyses

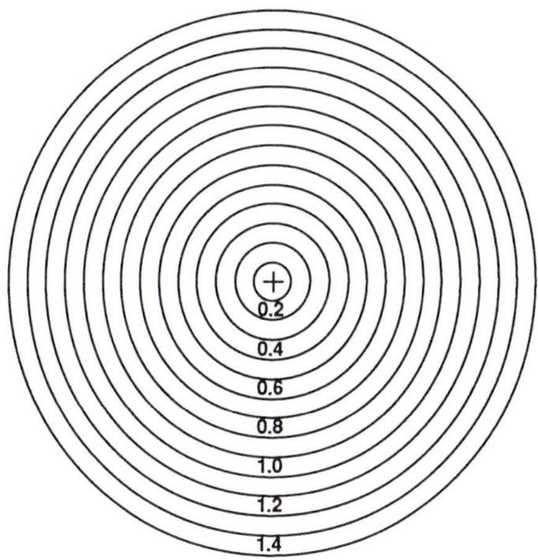

Figure 17-10. Standardized tolerance zone overlay

on a grid laid out to the same scale as the tolerance zone overlay. This is shown in Figure 17-9. Transparent tolerance zone overlays can be drawn as shown in Figure 17-10.

Inspection Procedure Using the Standardized Forms

Using the standard grid and tolerance zone overlays, the inspection process of Chapter 16 can be revised as follows:

1. Measure all feature sizes. If feature sizes are out of tolerance, the part can be rejected at this point.
2. Position the part and establish the datum for measurement.
3. Measure all feature locations and establish coordinate deviations. Plot the deviations on the grid and code them so as to identify the features which they repre sent.
4. Lay the tolerance zone overlay on the grid. If all the feature locations can be contained within the circle representing the specified position tolerance, the inspection is complete.
5. If the part is not acceptable at step 4, calculate the allowable position tolerance for each feature and mark the feature code number on the tolerance zone overlay at the appropriate circle.
6. Lay the tolerance zone overlay on the grid. If all of the feature locations can be contained within their respective tolerance zones, the part is acceptable and the inspection is complete.
7. If the part is not accepted at step 6, it may be appropriate to determine if the part can be reworked to meet specifications.

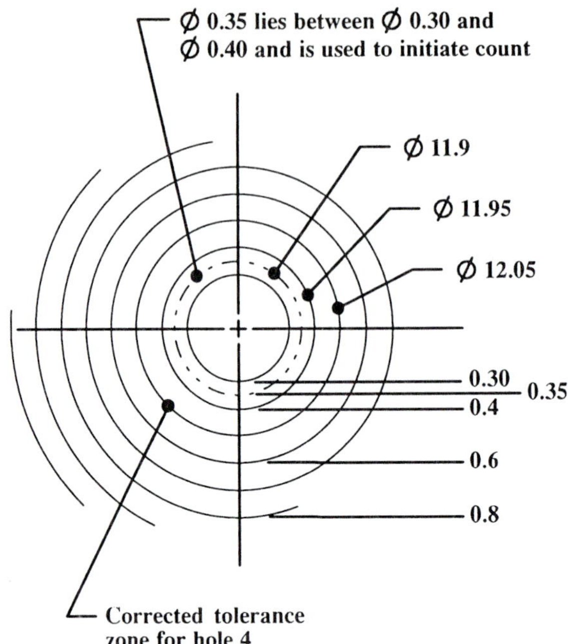

Figure 17-11. Graphical procedure for determining allowable tolerance

Graphical Procedure for Determining Allowable Tolerance

When it is necessary to determine allowable position tolerance, it may be calculated as in Example 17-3, or it may be approximated by the following procedure. First, find on the overlay the circle which represents the critical tolerance value. Given a value of ∅ 0.35, this would lie between ∅ 0.30 and ∅ 0.40 in Figure 17-11, since the tolerance zones are laid out in 0.10 diametral increments.

Starting from the circle representing the critical tolerance value, assume a corresponding feature size equal to the MMC (or other critical size as determined by the specification) feature size. Count outward on the radial lines until the measured feature size is reached. The final point represents the corrected tolerance zone. Figure 17-11 represents the determination of the allowable tolerance for Hole 4 of Example 17-1.

Example 17-1

This example shows the modified paper gage for the part inspected by conventional methods in Example 16-4. The measured data are repeated here for convenience. In that example, the part was easily accepted for pattern location requirements, but appeared to be unacceptable for within pattern requirements until we evaluated it by a trial and error iterative method. With the paper gage method, we are able to free the measured locations from the datums and the location grid and the tolerance zones are allowed to float relative to one another. Now we can see that the part is truely an acceptable part.

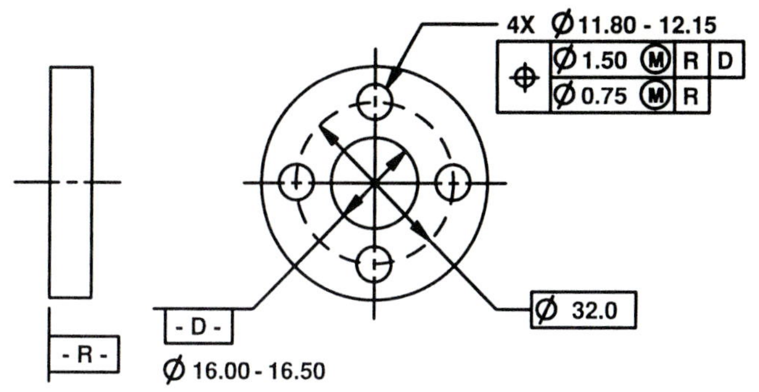

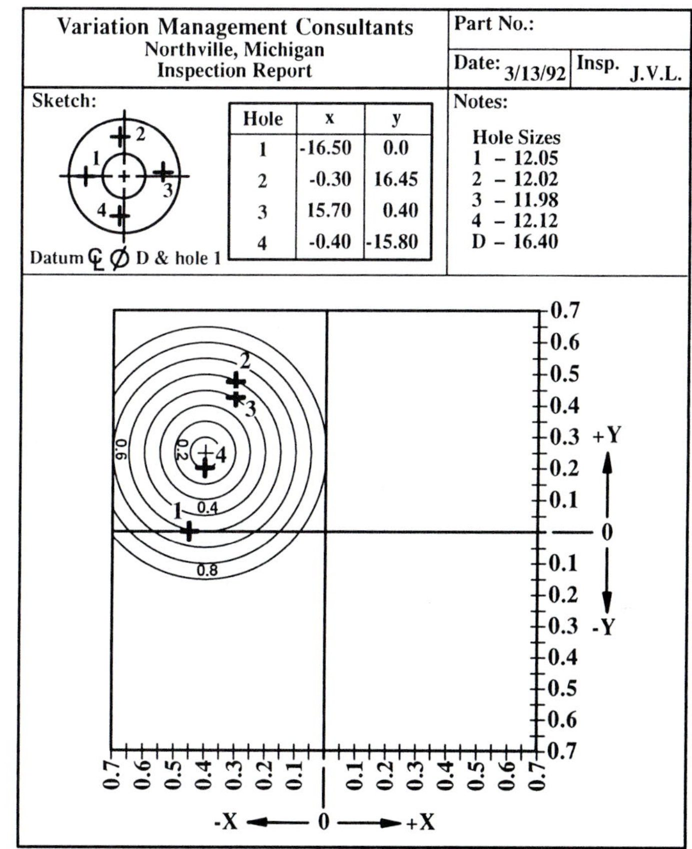

Variation Management Consultants
Northville, Michigan
Inspection Report

Part No.:

Date: 3/13/92 Insp. J.V.L.

Sketch:

Hole	x	y
1	-16.50	0.0
2	-0.30	16.45
3	15.70	0.40
4	-0.40	-15.80

Datum ℄ Ø D & hole 1

Notes:

Hole Sizes
1 – 12.05
2 – 12.02
3 – 11.98
4 – 12.12
D – 16.40

Example 17-1.

Example 17-2

Evaluate the inspection data shown by use of the modified paper gage technique.

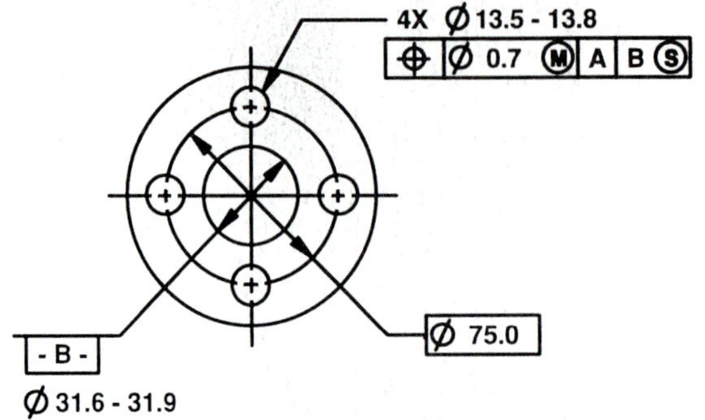

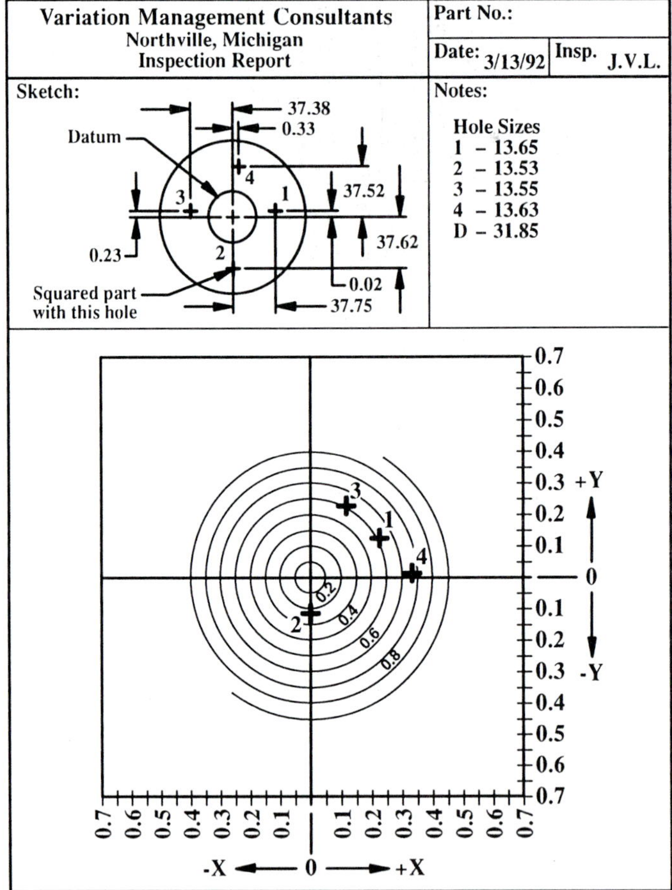

Solution The location deviations are determined and plotted on the grid. Again they are such that the allowable tolerance need not be determined because they lie within the zone of the critical tolerance value. Since the datum is stated to be critical at RFS size, the datum reference point on the grid and the overlay must be exactly aligned.

Example 17-3

Evaluate the data given by use of a modified paper gage.

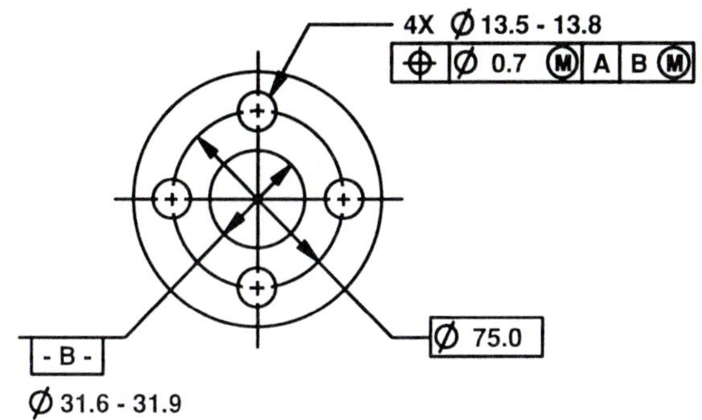

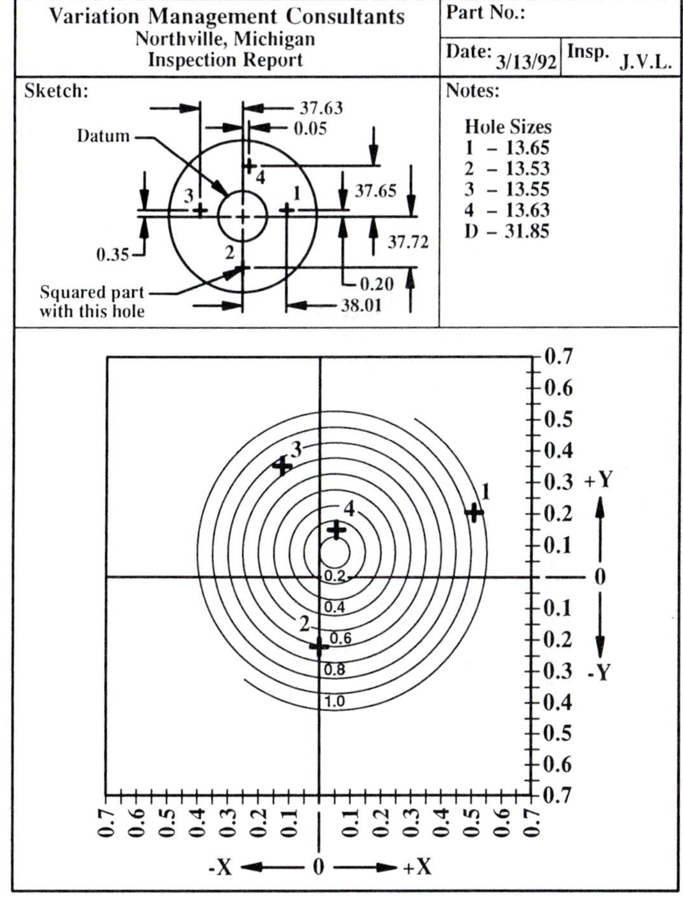

Solution This example is similar to Example 17-2 except that an MMC datum is specified and the reference point on the grid is therefore required to lie within a corresponding datum tolerance zone, which is calculated using the additional tolerance concept. The allowable tolerance zone sizes can easily be determined to be:

Datum	∅ 0.25	Hole 2	∅ 0.73	Hole 4	∅ 0.83
Hole 1	∅ 0.85	Hole 3	∅ 0.75		

Displacement of the tolerance zone grid from the center of the position tolerance overlay, but within the datum tolerance zone, is necessary to accept the part.

Example 17-4

This example, although it is admittedly a simple one, shows how the modified paper gage technique can be used to evaluate features that are out-of-square relative to the stated or implied datums. Because a zero MMC position tolerance is stated, the allowable tolerance at each hole must be determined. The points representing the top of the hole are labeled *T*, while those representing the bottom are labeled *B*. Obviously, both points for any one hole must lie within the proper tolerance zone for the part to be acceptable.

Solution The allowable tolerance zones are determined to be:

Hole 1 ∅ 0.35

Hole 2 ∅ 0.30

Hole 3 ∅ 0.20

Because the top and bottom of Hole 3 cannot be contained within its ∅ 0.20 tolerance zone, the part is rejected.

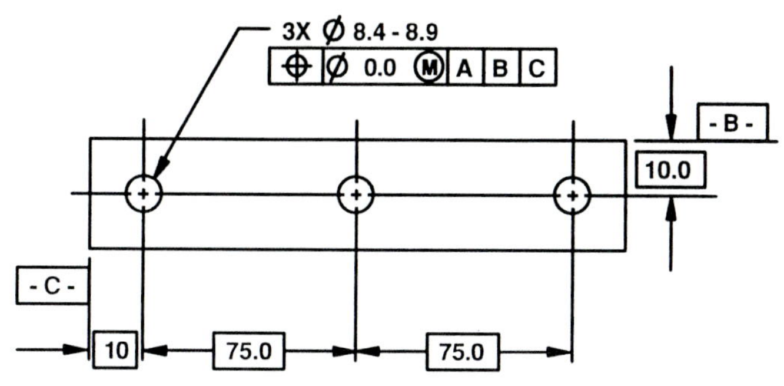

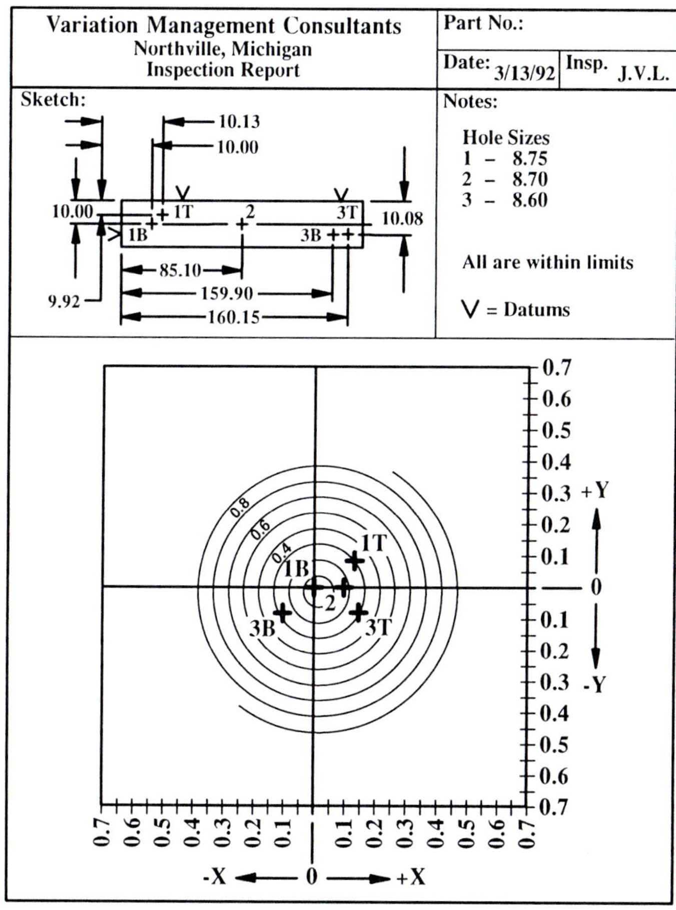
Variation Management Consultants
Northville, Michigan
Inspection Report

Part No.:
Date: 3/13/92
Insp. J.V.L.

Sketch:

Notes:

Hole Sizes
1 – 8.75
2 – 8.70
3 – 8.60

All are within limits

V = Datums

Example 17-5

Determine how to rework the part of Example 17-4 to meet product specifications.

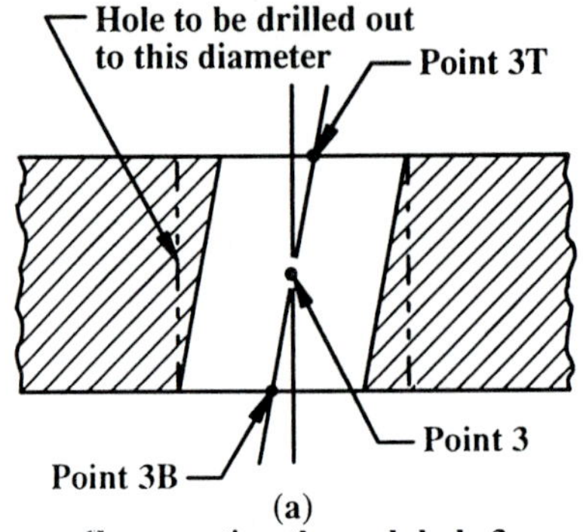

(a)
Cross section through hole 3

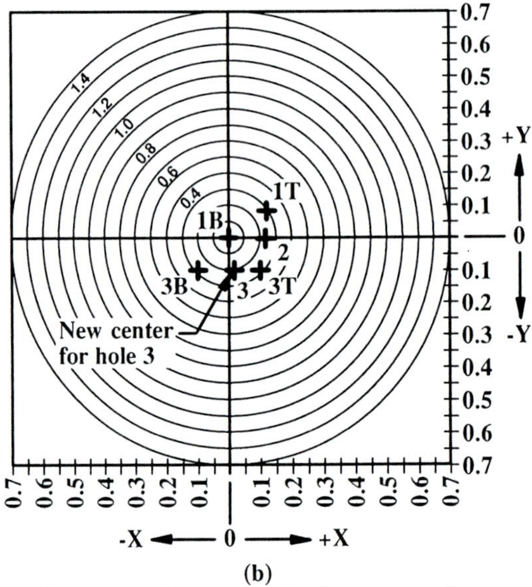

(b)
Paper gage for proposed salvage operation

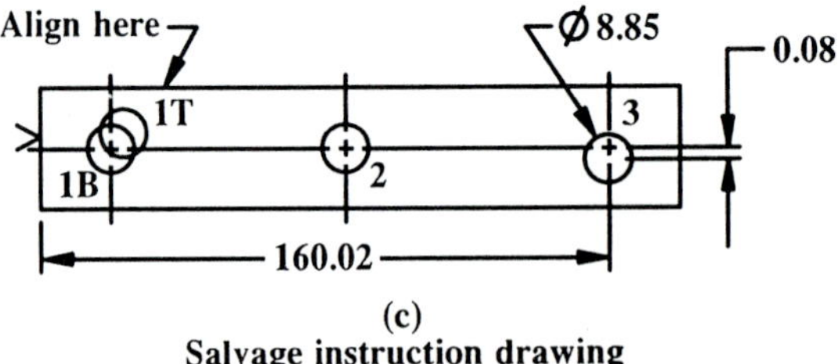

(c)
Salvage instruction drawing

Solution As shown in the illustration the smallest hole which would square up Hole 3 would be oversized by the spread between hole centers *3B* and *3T*, and its center would lie midway between these points or at approximately $X = 0.02$, $Y = -0.08$. The resulting hole size would be:

$$H_A = 8.60 + (0.10 + 0.15) = 8.85$$

Although near the high limit, this hole size is acceptable. Applying Equation 9-4:

$$T_3 = T_{ADD} = H_A - H = 8.85 - 8.4 = 0.45$$

The paper gage is replotted and the position is seen to be acceptable. The final illustration represents the instructions necessary to carry out the salvage operation.

Additional Applications

This technique can also be used to evaluate patterns of projections, non-circular features such as square holes, and RFS or LMC features as well. In each case, consideration must be given to the appropriate tolerance zone shape and the proper allowable position tolerance.

PROCESS CAPABILITY STUDIES

The modified paper gage can be used to study the capability of a process involving position tolerance control. This may be done for several reasons, for example, to analyze the data from a preproduction run, or to study a process which has a high rate of unacceptable parts.

The technique is rather simple. A number of parts are inspected and the feature locations are plotted on the tolerance grid. Then all the points representing a certain feature are bounded by a circle. The minimum size tolerance zone corresponding to each particular feature is chosen and the usual evaluation is made, the only difference being that the locations which are usually represented by a point are now represented by the area contained within a circle.

Other statistical methods help to define the results of using a small number of parts to represent the total production volume. For example, at a 99% confidence level, 50% of all production parts will lie within the predicted circle if 10 or more parts are sampled. If the number of samples is increased to 100, about 94% of all production parts will lie in the observed range. Since it is necessary for the process range to be less than the available tolerance to allow for drift, tool wear and the like, the circles representing the feature locations can be increased in size to allow a margin for such variables. Since this technique has not been widely used to date, factors are not readily available and judgment must be applied. It is obvious, however, that a large sample size is a better approach if at all possible. Most quality control handbooks define appropriate sample sizes.

Figure 17-12 illustrates the results of a hypothetical capability study.

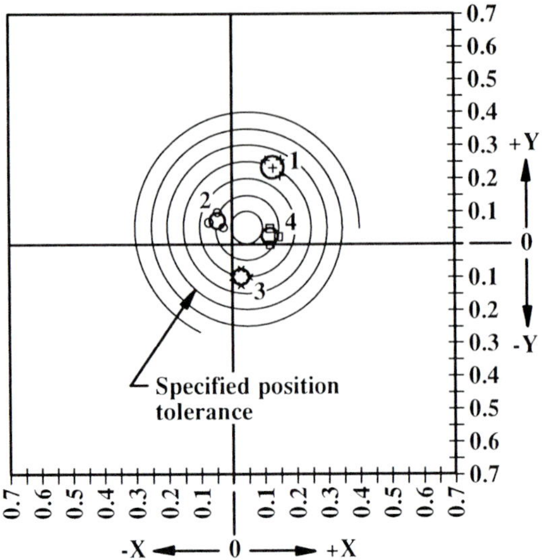

Figure 17-12. Results of a hypothetical process capability study

SUMMARY

Paper gaging is an inspection technique which allows data taken from an inspection to be evaluated in the same way that a composite functional gage verifies a part. It is fundamentally an enlarged layout of the tolerance zones and the measured feature locations.

The modified paper gage superimposes the feature locations and tolerance zones to common centers to reduce the layout size. The resulting increase in the size of the laid out deviations reduces error due to inaccuracy in the layout work. Standardized position grids and tolerance zone overlays can be developed for the modified paper gage. The position of the component during measurement requires careful consideration since it can introduce error when the modified paper gage is used. The modified paper gage can be used to conduct process capability studies.

Although to the author's knowledge, paper gaging techniques have not been applied via computer simulation methods, it seems to be an obvious method for the evaluation of inspection data.

Appendix 1

Glossary

Inasmuch as the purpose of geometric dimensioning and tolerancing is to aid in analyzing and documenting component functional requirements, it is necessary that it be based upon rigorously defined terms. A familiarity with these terms should be attained. Abbreviations and symbols used in the text are shown here following appropriate definition headings. A separate listing of symbols in alphabetical order is provided in the front of the book.

DEFINITIONS

Balance Dimension. The sum or difference of two machining or balance dimensions used for a process chart.

Balanced Position Tolerance. See position Tolerance, Balanced.

Benderize. An expansion factor applied to a statistical stack-up to compensate for the lack of conformance to statistical assumptions; specifically being on target.

Bilateral Tolerance. Equal plus and minus tolerance values.

Capable Process. A process under statistical control (i.e., all its variation is from Common Causes).

Centering Index (C_C). The ratio of the error in location of the process mean divided by one half of the tolerance range.

Class I and II Errors. A class I error rejects a good part, while a class II error accepts a bad part.

Clearance (CL). The total space between mating parts.

Common Cause. A source of variation that is always present, part of the random variation inherent in the process itself. Its origin can usually be traced to an element of the system which only management can correct.

Concentricity (*C*). Eccentricity measured as a total value. Concentricity equals two times eccentricity.

Concentricity, Critical (C_C). The minimum allowable concentricity.

Coordinate Tolerance Zone. See Tolerance Zone, Coordinate.

Critical position Tolerance. See position Tolerance, Critical

Critical Size. The size within a tolerance range that is most critical relative to the function of a feature of size.

Datum. A specifically determined origin from which dimensions are expressed and measurements are taken.

Datum Tolerance. See Tolerance, Datum.

Eccentricity (*E*). The amount or state of being off center.

Expected Variation. The portion of an RMS result that is free of fixturing (centering) errors.

Feature. A specific characteristic or component portion of a part, such as a hole, slot, boss, and so on.

Fixed Fastener. A fastener rigidly mounted in one part and passing through clearance holes in subsequent parts (i.e., a bolt mounted into a tapped hole, a dowel press fit into a hole, and the like).

Floating Fastener. A fastener passing through clearance holes in two or more parts.

Frequency Distribution. A plot of occurrences versus some parameter such as dimensional size.

Gilson Factor. Another expansion factor for stack up calculations. It produces results similar to the Bender method.

Histogram. A bar graph of data.

Hole Pattern. Any group of holes which function as one unit regardless of differences in size, plane of location, and so on.

Independent. No interaction between separate parameters, when used in statistical calculations.

Interchangeability. A principle of design wherein universal exchange of mating parts is possible. Parts assemble and function properly without machining or fitting at assembly.

Least Material Condition (LMC). The critical size for parts which require positioning so that they will assemble properly. An LMC feature is one that is derived from its toleranced features so that it contains the least amount of material with respect to any portion of the feature which will affect a position or form tolerance. For a hole, LMC represents the high-dimensional limit. For an outside diameter, LMC represents the low-dimensional limit.

Limit Dimensions. Dimensions that reflect the minimum and maximum size of a component.

Line of Orientation. A line that determines the radial orientation of other features.

Loss Function. A plot of a characteristic, such as a dimension, and the loss in value that occurs as the characteristic varies from a customer based target or nominal value.

Maximum Material Condition (MMC). The critical feature size for parts which require clearance to assemble. An MMC feature is one that is derived from its toleranced dimensions so that it contains the maximum amount of material with respect to any portion of the feature which will affect a position or form tolerance. For a hole, MMC represents the low-dimensional limit. For an outside diameter, MMC represents the high-dimensional limit.

Monte Carlo Simulation (MCS). A computer-based procedure that simulates the sampling of known or assumed statistical distributions.

Natural Tolerance. Product variation.

Normal Distribution. The classic bell shaped symmetrical frequency distribution.

Position Tolerance. Limits within which a feature of size may be displaced from its true position.

Position Tolerance, Balanced (T_B). A distribution of the total position tolerance equally between mating parts.

Position Tolerance, Critical (T_C). The minimum allowable position tolerance.

Position Tolerance, Total (T_T). The sum of the position tolerance for two mating parts.

Position Tolerance, Unbalanced. A distribution of the total position tolerance unequally between mating parts.

Projected Tolerance Zone. See Tolerance Zone, Projected.

Process capability Ratio (C_p). The ratio of product tolerance divided by process spread.

Process centering Ratio (C_{pk}). The ratio of the distance from the process mean to the nearest product tolerance limit divided by one half of the process spread.

Rectangular Distribution. A frequency distribution for processes or events with all outcomes having equal probability.

Regardless of Feature Size (RFS). Occupies a middle ground between MMC and LMC. It is associated with functions where alignment occurs, as in the case of press fit components, RFS indicates that a feature is always at its critical size, hence it can be conveniently thought of as indicating the center line of a feature.

Resultant Size (R). The effective size of a feature that must be considered when determining the effects of position and size tolerance cumulation.

Root Sum Square (RSS). A statistical stack- up method that assumes that all components and the resulting assembly have normal distributions.

Selective Assembly. A method of assembling sorted components to reduce assembly variation.

Shimmed Assembly. An assembly method which uses shims to reduce assembly variation.

Skewed Distribution. A non-symmetrical frequency distribution with the majority of observations biased toward one side or the other.

Special Cause. An intermittent source of variation that is unpredictable, or unstable; sometimes called an assignable cause. It is signaled by a point beyond the control limits or a run or other non-random pattern of points within the control limits.

Statistical Process Control (SPC). A statistical method used to identify Common Cause variation of a process and to use this data for subsequent process control or process capability analysis.

Tolerance. The permissible variation in a specified relationship, such as Form, Attitude, or Location.

Tolerance, Datum. A tolerance describing the limits within which a datum feature may be displaced from it true position.

Tolerance Zone, Coordinate. A square or rectangular tolerance zone derived from the tolerances of coordinate dimensions.

Tolerance Zone, Position. The zone about true position which limits the position of a feature.

Tolerance Zones, Projected. A tolerance zone projected above a datum surface describing the limits within which the extended center line of a feature may be displaced from true position.

True Position. The theoretically exact location of a feature.

Unbalanced Tolerance. Tolerances that have different values in opposite directions from the nominal value.

Unbalanced position Tolerance. See position Tolerance, Unbalanced.

Unilateral Tolerance. A tolerance that applies only in one direction.

Variation. The change that occurs in size or shape of features or parts. Variation can occur within a piece, from piece to piece, and over time. Variation may be measured as a piece to piece range or relative to a desired nominal value.

Virtual Condition (VC). A composite of another critical size, typically MMC, and another tolerance such as Perpendicularity, or Position.

Virtual Fastener Size (F_V). The size of a fastener when its critical size is modified to account for another tolerance, often allowable eccentricity.

Virtual Hole Size (H_V). The size of a hole formed by assembling two or more parts, or by considering its critical size and another tolerance.

Appendix 2

Z Table

Z Table

Standard Normal Probability Distribution

Areas under the Standard Normal Probability Distribution

between the Mean and Successive Value of z.*

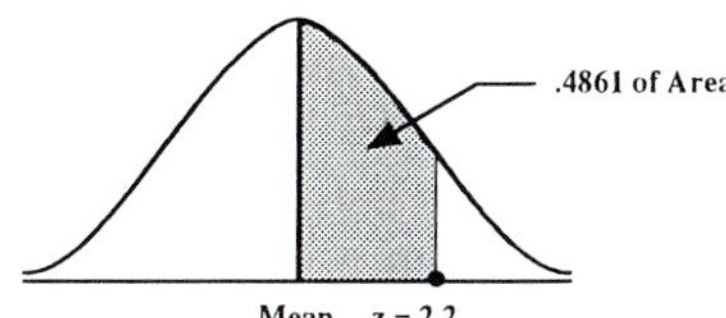

Example: To find the area under the curve between the mean and a point 2.2 standard deviations to the right of the mean, look up the value opposite 2.2 in the table; .4861 of the area under the curve lies between the mean and a z value 2.2.

Z	.00	.01	.02	.03	.04	.05	.06	.07	.08	.09
0.0	.0000	.0040	.0080	.0120	.0160	.0199	.0039	.0279	.0319	.0359
0.1	.0398	.0438	.0478	.0517	.0557	.0596	.0636	.0675	.0714	.0753
0.2	.0793	.0832	.0871	.0910	.0948	.0987	.1026	.1064	.1103	.1141
0.3	.1179	.1217	.1255	.1293	.1331	.1368	.1406	.1443	.1480	.1517
0.4	.1554	.1591	.1628	.1664	.1700	.1736	.1772	.1808	.1844	.1879
0.5	.1915	.1950	.1985	.2019	.2054	.2088	.2123	.2157	.2190	.2224
0.6	.2257	.2291	.2324	.2357	.2389	.2422	.2454	.2486	.2517	.2549
0.7	.2580	.2611	.2642	.2673	.2704	.2734	.2764	.2794	.2823	.2852
0.8	.2881	.2910	.2939	.2967	.2995	.3023	.3051	.3078	.3106	.3133
0.9	.3159	.3186	.3212	.3238	.3264	.3289	.3315	.3340	.3365	.3389
1.0	.3413	.3438	.3461	.3485	.3508	.3531	.3554	.3577	.3599	.3621
1.1	.3643	.3665	.3686	.3708	.3729	.3749	.3770	.3790	.3810	.3830
1.2	.3849	.3869	.3888	.3907	.3925	.3944	.3962	.3980	.3997	.4015
1.3	.4032	.4049	.4066	.4082	.4099	.4115	.4131	.4147	.4162	.4177
1.4	.4192	.4207	.4222	.4236	.4251	.4265	.4279	.4292	.4306	.4319
1.5	.4332	.4345	.4357	.4370	.4382	.4394	.4406	.4418	.4429	.4441
1.6	.4452	.4463	.4474	.4484	.4495	.4505	.4515	.4525	.4535	.4545
1.7	.4554	.4564	.4573	.4582	.4591	.4599	.4608	.4616	.4625	.4633
1.8	.4641	.4649	.4656	.4664	.4671	.4678	.4686	.4693	.4699	.4706
1.9	.4713	.4719	.4726	.4732	.4738	.4744	.4750	.4756	.4761	.4767
2.0	.4772	.4778	.4783	.4788	.4793	.4798	.4803	.4808	.4812	.4817
2.1	.4821	.4826	.4830	.4834	.4838	.4842	.4846	.4850	.4854	.4857
2.2	.4861	.4864	.4868	.4871	.4875	.4878	.4881	.4884	.4887	.4890
2.3	.4893	.4896	.4898	.4901	.4904	.4906	.4909	.4911	.4913	.4916
2.4	.4918	.4920	.4922	.4925	.4927	.4929	.4931	.4932	.4934	.4936
2.5	.4938	.4940	.4941	.4943	.4945	.4946	.4948	.4949	.4951	.4952
2.6	.4953	.4955	.4956	.4957	.4959	.4960	.4961	.4962	.4963	.4964
2.7	.4965	.4966	.4967	.4968	.4969	.4970	.4971	.4972	.4973	.4974
2.8	.4974	.4975	.4976	.4977	.4977	.4978	.4979	.4979	.4980	.4981
2.9	.4981	.4982	.4982	.4982	.4984	.4984	.4985	.4985	.4986	.4986
3.0	.4987	.4987	.4987	.4988	.4988	.4989	.4989	.4989	.4990	.4990

*From Robert D. Mason, *Essentials of Statistics*, ©1976, p. 307. Adapted by permission of Prentice Hall, Englewood Cliffs, N.J.

Appendix 3

$\overline{X}$ AND S CHARTS

The calculation of the statistics used in the X and S charts are based upon:

n = number of samples in a subgroup

k = number of subgroups

$\overline{X} = \Sigma X_i / n = (X_1 + X_2 + \ldots + X_n)/n$

$\overline{\overline{X}} = \Sigma X_i / k = (\overline{X}_1 + X_2 + \ldots + X_k)/k$

$UCL_X = \overline{\overline{X}} + A_3 S$

$LCL_X = \overline{\overline{X}} - A_3 S$

$S = \sqrt{\Sigma(X_i - \overline{X})^2)/n - 1}$

$S = S_i / k$

$UCL_S = B_4 \overline{S}$

$LCL_S = B_3 \overline{S}$

where:	n	A_3	B_3	B_4
	2	2.659	—	3.267
	3	1.954	—	2.568
	4	1.628	—	2.266
	5	1.427	—	2.089
	6	1.287	0.030	1.970
	7	1.182	0.118	1.882
	8	1.099	0.185	1.815
	9	1.032	0.239	1.761

Appendix 4

Computer Programs

Name/Source	*Type*
SPC Plus Advanced Systems and Designs 27200 Haggerty Road Farmington Hills, MI 48331	A comprehensive, menu driven SPC program for both variable and attribute control charts.
Statgraphics STSC Inc. 2115 East Jefferson Street Rockville, MD 20852	Statistical visualization software for total data analysis.
Statistical Analysis for Engineers J. Wesley Barnes Prentice Hall, Inc. Englewood Cliffs, NJ 07632	A statistics text with supporting computer programs.
Toltech System The Norwegian Institute of Tech. Rich. Birkeiandsv, 2B N-7034 Trondheim, Norway	Statistical based tolerance analysis with cost based tolerance distribution consideration.
VSA Applied Computer Solutions 300 Maple Park Blvd St. Clair Shores, MI 48080	A Monte Carlo-based statistical program to simulate the variation of industrial products.
VSM/PC EDS Division of General Motors Southfield, MI	An automotive oriented, Monte Carlo program with availability limited to General Motors divisions and to suppliers.

Index

A

B

C

D

F

G

H

I

K

L

M

N

P

R

S

T

V

W